土木工程、环境工程和材料科学的创新、应用和教育

Innovation, Application and Education on Civil Engineering, Environment Engineering and Material Science

主 编 刘俊伟 苗吉军 吕 平 ［德］葛莲·盖克 ［德］雷纳·蒙塞斯

Chief Editor: Liu Junwei, Miao Jijun, Lyu Ping,
Gilian Gerke, Rainer Monsees

大连理工大学出版社

图书在版编目(CIP)数据

土木工程、环境工程和材料科学的创新、应用和教育＝Innovation, Application and Education on Civil Engineering, Environment Engineering and Material Science / 刘俊伟等主编. -- 大连：大连理工大学出版社，2022.6

ISBN 978-7-5685-3023-1

Ⅰ.①土… Ⅱ.①刘… Ⅲ.①土木工程－国际化－人才培养－研究②环境工程－国际化－人才培养－研究③工程材料－国际化－人才培养－研究 Ⅳ.①TU②X5③TB3

中国版本图书馆 CIP 数据核字(2021)第 097424 号

大连理工大学出版社出版

地址：大连市软件园路 80 号　邮政编码：116023
发行：0411-84708842　邮购：0411-84708943　传真：0411-84701466
E-mail：dutp@dutp.cn　URL：http://dutp.dlut.edu.cn
大连图腾彩色印刷有限公司印刷　　大连理工大学出版社发行

幅面尺寸：210mm×285mm　印张：13.5　字数：308 千字
2022 年 6 月第 1 版　　2022 年 6 月第 1 次印刷

责任编辑：王晓历　　　　　　　　责任校对：白　露
　　　　　　　　　　封面设计：张　莹

ISBN 978-7-5685-3023-1　　　　　　定　价：69.00 元

本书如有印装质量问题，请与我社发行部联系更换。

前 言
PREFACE

青岛理工大学历来重视国际合作交流,与国际 90 余所知名高校建立了校际交流和合作关系。本书是第一届青岛理工大学-马格德堡应用技术大学中德学生会议(1st QUT-HSMD International Student Conference)的会议论文集。

中德学生会议由青岛理工大学土木工程学院承办,召集了中德学生和学术团体,共同分享、交流各自领域的创新思路与成果,对参会者的学业和文化交流起到巨大的促进作用。本书收录了会议中分享人才培养探索与实践、土木工程、环境工程与资源综合利用、材料科学与工程等四大方面的新成果。

青岛理工大学与马格德堡应用技术大学有着长久的合作历史,未来的合作也将延绵不绝,中德学生系列会议也已成为全世界的学生论坛,为更多的中德学生和世界其他各国学生提供交流的平台。

编 者
2022 年 6 月

所有意见和建议请发往:dutpbk@163.com
欢迎访问高教数字化服务平台:http://hep.dutpbook.com
联系电话:0411-84708462 84708445

目 录
CONTENTS

1. Research on Teaching Technology and Practice ············· 1

 (1) Capacity Building—Global Challenges and Chances of International Teaching and Learning Exchange ············· 3

 (2) Internationalization of Talent Training in Civil Engineering and Material Science ············· 8

 (3) From Students, for Students, Let's Build the Future ············· 17

 (4) Research on the Training Mode of Applied Innovation Talents in the Major of Material Science Engineering-take Qingdao University of Technology as an Example ············· 21

2. Civil Engineering ············· 29

 (1) Fire Protection Engineering Preventive Structural Fire Protection and Engineering Methods in Germany ············· 31

 (2) Construction and Design of Skyscrapers ············· 34

 (3) Surprises During Remediations ············· 38

 (4) Supporting an Excavation Pit with Steel Piling ············· 41

 (5) How to Make a High-speed Railway Tunnel Safe in Case of Emergency ············· 44

 (6) Research on TMD Parameters for Swing Vibration Control of Suspended Structures ············· 49

 (7) Research on the Vibration Response of Subway with Disk-spring and Thick Rubber Isolator ············· 54

 (8) Carbon Source Supplement Strategies Make Difference in a Semi-centralized Black Water Treatment System ············· 61

 (9) Influence of Reinforcing Ring Parameters on Seismic Performance of Rectangular Steel Tubular Column-H Shaped Steel Beam Joint ············· 65

(10) Study on Seismic Behavior of Steel Structure Joints with Inner Diaphragms and Outer Ring Plates ……………………………………………………………… 70

(11) The Characteristics of Underground Comprehensive Pipe Gallery Structure and Waterproof Practices ………………………………………………………… 75

(12) Summary of Research on Pile-soil Dynamic Response under Lateral Cyclic Loading ……………………………………………………………………………… 80

(13) Preliminary Study on Laboratory Model Test under Vertical Cyclic Loading … ……………………………………………………………………………… 85

(14) The Influence of the Number of Slots on the Damping Performance of the Stand-off Layer Damping Structure …………………………………………………… 91

3. Environmental Engineering and Comprehensive Utilization of Resources …………… 97

(1) Rewetting of Fens: a Case Study from Northern Germany …………………… 99

(2) Formation and Succession of Oxbow Lakes in the Middle Elbe Biosphere Reserve ……………………………………………………………………………… 104

(3) The Elution of Heavy Metals at Tailings Piles ………………………………… 109

(4) Electronic Waste—Challenges and Chances ……………………………………… 111

(5) Urban Mining—Discovering the Values of Construction Products …………… 114

(6) The Elimination of Phosphorus on Sewage Treatment Plants ………………… 117

(7) Flood Protection Measures ……………………………………………………… 121

(8) Long-term Study for the Waste Behaviour of Different Population Groups in Germany ……………………………………………………………………… 124

(9) Sorting Analysis and Material Testing of Plastic Waste from the Sea ……… 129

(10) Decentralized Rainwater Management …………………………………………… 133

4. Materials Science and Engineering ………………………………………………… 137

(1) Structures Made of Waterproof Concrete ……………………………………… 139

(2) Research on Resistance to Environmental Corrosion for the Qingdao Bay Bridge Protection ……………………………………………………………………… 143

(3) Preparation and Characterization of Fe_3O_4/P (MAA-EGDMA)/Au Catalyst ········ 149

(4) Backward Response Law of Relative Humidity of Unsaturated Concrete Under General Atmosphere Environment in Qingdao Area ·············· 153

(5) Molecular Dynamics Study on the Adsorption Properties of Ions at the Surface of Sodium Alumino-silicate Hydrate (NASH) Gel ·············· 158

(6) Synthesis and Properties of Phosphorus Flame Retardant Polyurea Elastomers ················ 162

(7) Effect of Curing Regimes on Short and Long-term Compressive Strength Development of Geopolymer Concrete ·············· 167

(8) Review on Research on Epoxy Resin Modified Repair Mortar Based on Alkali-activated Cementitious Materials ·············· 173

(9) The Preparation and Properties of a Novel Modified Flame Retardant Polyurea Coating ·············· 178

(10) Preparation and Catalytic Performance of Magnetic Core Shell Fe_3O_4/P(VP-EGDMA)/Ag ·············· 183

(11) Preparation and Properties of Halogen Flame Retardant Polyurea Elastomers ················ 188

(12) Research Progress on Measurement and Calculation Methods of Damping Material Loss Factor ·············· 193

(13) Research Progress on Viscoelastic Damping Materials and Damping Structures ················ 198

(14) Water Capillary Absorption of Alkali Slag Concrete after Salt-frost Action ·············· 204

1. Research on Teaching Technology and Practice

(1)Capacity Building—Global Challenges and Chances of International Teaching and Learning Exchange

Prof. Dr. G. Gerke, Prof. Dipl.-Ing R. Monsees

Technical University Magdeburg/Stendal, University of Applied Sciences

Department of Water, Environment, Civil Engineering and Safety

Abstract

Since ancient times, people have learned to gain knowledge and to impart it to others, first via a spoken word and then in a written form. The target was always to teach others and to secure the knowledge for the future. This was the foundation for the development of our society. Over the years due to the rapid development of technologies and innovations in different spheres of society, it has dramatically changed the structures of communications and multicultural exchanges, as well as the ways they interact and function. Even though gaining knowledge from paper-based materials and attending monologue-based lectures were the only existing method, what we can observe nowadays is that the methodology of teaching and the ways to learn something new need other techniques and mechanisms. To reach young people in a world of fast Internet and social media is a quick changing challenge for didactics. Teaching is more and more like coaching, motivation and moderation. The target is to prepare young people for a global, international, fast changing work market and to transfer knowledge in a sustainable way. Next to all information raining on to students, the expert-knowledge needs to be anchored. Here new and matching didactic methods are required. The following text gives a short overview of the student conference as a possibility to link students from two countries or more if needed, and provide a platform for an international exchange.

Key Words

Heavy Metals, Tailings Piles, Elution

1 Motivation

With the possibilities of modern technology the world gets smaller and distances are shorter. To contact other countries is so easier nowadays and projects are easy to be handled over far away borders. Communication is a question only of hours, no matter how far the partner is away.

These facts give a new need to education in schools and universities. Next to professional competences, other competences have to be part of a modern curriculum within a study program. These are for example social competence, team ability, languages or method, action and media competences.

Students have to prepare during their studies for the future working environment. This needs an adaption within the didactic methods of teaching. Students should gain as much experiences already during their time at the university. This needs motivation.

An important didactical characteristic to support the motivation of students is to integrate reflexion time. This gains learning experiences and teaches to take on responsibility for own actions. This also strengthens social alliances [1].

To connect learning with research as soon as possible, research should be integrated in lectures in the form of projects. So students can get used to project work, project management and gain the ability of taking decisions and responsibility. These key competences are one of the most important factors in the later environment of working especially being in an executive position.

On an international level a term of "Teaching-Research-Nexus" (TRN) can be found. This shows a relation between both academic activities and is good for both sides. Both key aspects get the same importance within the academic training[2,3]. The target is on the base of integrated thinking to provide students with the competence of being able to take decisions and then have the will to do so. This is important and has to be transferred into the later practice[4].

The given student conference is a perfect example of integrated research and teaching into the curriculum and is a connection not only of different student programs but also a bridge to another culture and country. This is an example of teaching international competence in the field of Teaching-Research-Nexus.

2 The Student Conference

A couple of years ago, the School of Civil Engineering within QUT and the Department of Civil Engineering at the University of Applied Sciences had held a bilateral cooperation where students and personnel had an opportunity to take part in different activities as excursions, visiting partner university, 'semester abroad' study program, cooperative supervisions of theses, guest lectures and plenty of other bilateral activities were organized.

Figure 1　Visit of Prof. Gilian Gerke at the campus of TU Qingdao, 2017

The overall target is to establish a strategic partnership within three phases: Teaching-Research-Further Education. According to these phases after building the foundation and being integrated, students are brought into the position to gain experiences in an international frame.

One of the formats of the academic work is a conference. This is the platform to gather experts and to support professional exchange. Normally, students don't get into the position of being part of a conference during their time at the university. We provide them with a chance of a working environment and professional support. Students should gain their competences also for these situations. This is the idea of the 1st QUT-HSMD International Student Conference. It's a student-to-student conference where everything is laid into the hands of students. They are in charge of the organisation and the execution of the conference. On both sides of China and Germany, an organisation team has to take over the responsibility for the whole conference and the framework program. Conference presentations and session moderations are performed by students. This makes the conference special. The role of the professors is motivation, coaching and consulting. Conference language is English, so all the students of both countries are on the same base line of communication.

Participating students have to apply with a short description of motivation and subject they want to present on the conference. This is already the first step to bring the students into the position of expressing themselves in an academic way. One of the most important parts for the creative directors of the project is to give students of all study programs and grades a chance to join. This strengthens the awareness for other subjects and fellow students, who normally would not meet during their studies. Furthermore, the multidisciplinarity is encouraged.

3 Lessons to Be Learned

The main focus for the creative directors of the project is to encourage young people to rise to the challenge of giving a presentation in English in a foreign country. Furthermore, they should take responsibility as a chance to create a project following their own ideas, needs and questions. To organise such a conference means a great gain of experiences in the field of international cooperation work and can strengthen the feasibility of own contribution and self-development. This is next to the professional knowledge and it is not a payable benefit during the time of studying. For normal students duties infill a semester nearly 100%, so they have no time to develop other important competences.

A positive feedback and a successful project is a motivating impact for the further steps a student takes in his or her studies. They also learn on the base of self-dependence to organize and perform a project. This strengthens the self-confidence. The interdisciplinarity work gains a wider horizon and gives self assurance within an international world. Besides all the professional facets, one very important aspect has also to be named. Here two cultures have the chance to come together and learn from each other. Both countries, China and Germany belong to the leading countries contributing the economical prosperity of the world. The conference itself and time spent in another country bring the possibility of getting to know another culture and learning beyond the own comfort zone.

Last but not least, for the coaching side, to accompany the students brings an aid on the way of the didactic abilities of the stuff and professors. The target here is to change rusty structures and to learn to keep an open mind for new teaching ideas and ways. They also have to learn to stay back and to let the students make their own experiences. That gains special skills for professors and changes the own perspective. This as a result brings students and professors into a win-win situation.

4 Conclusions

1st QUT-HSMD International Student Conference is an example to integrate teaching and research on an international base into the curriculum of students. In a global working field, it is important to introduce students as soon as possible to special skills besides the common knowledge of the field of study. The conference combines key competences like teamwork, language skills, international abilities, communication, project management and social qualifications with the professional knowledge that students are educated in their speciality study program. Following the saying: "learning by doing", the conference is going to become a platform to encourage students to take over responsibility for decisions and practice within a protected environment. This is part of the training due to the future practical working field. Besides all these aspects, an excursion like this widens the horizon of all participants-students, staff and professors and builds a step for a lifelong learning. Such a conference with all included experiences stays as a memory stronger than any other lecture in the curriculum.

References

[1] Gerholz, K.-H., Liszt, V.; Klingsieck, K. B. (2015). Didaktische Gestaltung von Service Learning - Ergebnisse einer Mixed Methods-Studie aus der Domäne der Wirtschaftswissenschaften. bwp @ Berufs- und Wirtschaftspädagogik - online, Ausgabe 28, 1-23.

[2] Boyd, W. E.; O'Reilly, M.; Buch-

er, D. ; Fisher, K. ; Morton, A. ; Harrison, P. L. ; Nuske, E. ; Coyle, Rebecca; R. ; Rendall, K. (2010). Activating the Teaching-Research-Nexus in Smaller Universities: Case Studies Highlighting Diversity of Practice, Journal of University Teaching & Learning Practice, 7(2), 2010.

[3] Magnell, M. ; Söderlind, J. ; Geschwind, L. (2016). Teaching-Research-Nexus in Engineering Education, Proceedings of the 12th International CDIO Conference, Turku University of Applied Sciences, Turku, Finland, June 12-16, 2016. Aufgerufen am 27. 02. 2017 von www. cdio. org/files/document/cdio2016/68/68_Paper_PDF. pdf.

[4] Locke, W. (2009). Reconnecting the Research-Policy-Practice Nexus in Higher Education: 'Evidence Policy' in Practice in National and International Contexts. High Education Policy (2009) 22: 119. doi: 10. 1057/hep. 2008. 3.

(2) Internationalization of Talent Training in Civil Engineering and Material Science

Lyu Ping, Miao Jijun, Liu Junwei, Sun Yujie
School of Civil Engineering, Qingdao University of Technology

Abstract

The implementation of initiative "Belt and Road" poses an urgent need for internationalized talents in civil engineering and material science and engineering. This paper introduces the overall objectives, teaching concepts, talent training ideas and teaching reform of School of Civil Engineering Qingdao University of Technology in recent years. The practice of internationalized talent training which integrates teaching in class and off class, in school and out of school, at home and abroad has greatly promoted the quality of talent training. Besides, with the Sino-US joint program in Civil Engineering, joint program by school and enterprises, overseas projects (groups) and internationalization of professional information, curriculum innovation and internationalization of faculty, we have greatly quickened the step of internationalization. Thus, the process of the internationalized talent training in civil engineering, material science and engineering has been improved in an all-round way.

Key Words

Internationalized Talent Training, International Education System, Internationalization of Teachers, Civil Engineering, Material Science and Engineering

Introduction

The construction of the international high-speed rail, large deep-sea port and cross-sea bridge along the "Belt and Road" poses an urgent need for a large number of international talents with a sense of social responsibility, proficiency in foreign languages, familiarity with international rules, international vision, a good command of opportunities in the global competition of civil engineering and material science and engineering. This paper introduces the practices of Civil Engineering College of Qingdao University of Technology in internationalization in recent years. The initiative is an important conception of China's opening to the outside world in the new period, which provides a significant strategic opportunity for further promoting the internationalization of higher education in China. How to improve the educational mode and method under the guidance of the "Belt and Road" national initiative and cultivate the internationalized talents in the new era is undoubtedly an important responsibility for educators in the future. The practice in international talent training mode, innovative curriculum system and internationalization of faculty aims to promote internationalization of the discipline of civil engineering, material science and engineering, and comprehensively enhance the internationalization of talent training.

1 Construction of International Educational System

1.1 Current Situation of Discipline and Specialty

Civil engineering is a traditional and advantageous specialty of our university, including national specialty, Shandong Brand specialty and Taishan scholar specialty. In 2009, it passed the evaluation of undergraduate specialty education of the Ministry of Housing and Construction. In 2011, it was approved as a pilot specialty of "Excellent Engineer Education Training Plan", and in 2014, it passed the evaluation of Undergraduate Specialty Education of the Ministry of Housing and Construction. Its validity period is 6 years. After years of development, this specialty has become one of the important bases for the training of senior talents, scientific research and new technology development in the field of civil construction in China. Material science and engineering specialty is the "Characteristic Specialty in Shandong", "Education and Training Plan for Excellent Engineers" of the Ministry of Education, and the "Key Construction Specialty of Construction Engineering of Famous Universities with Shandong Characteristics". It is the base of Innovation and Intelligence Introduction of Marine Environment Concrete Technology, the Engineering Technology Center of the Ministry of Marine Environment Concrete Technology Education, the Key Concrete Laboratory of Universities in Shandong Province, and the Concrete Construction Specialty of Shandong Province, Research Center of Structural Durability Engineering Technology. The specialties of Civil Engineering and Material Science and Engineering rely on strong discipline advantages to train high-level international talents.

1.2 Overall Goal of Specialty Construction

In view of the new strategic demand for innovative talents in civil engineering and material application in our country and region, in accordance with the characteristics and advantages of civil engineering and material science and engineering specialty of our college, we try to carry out specialty construction according to local conditions. On this basis, the personnel training mode and the corresponding curriculum system should be reformed profoundly with the times, with the training of "applied" innovative talents as the orientation, and the application-oriented senior professional and technical personnel with innovative ability, strong engineering practice ability, international perspective and international competitiveness should be trained.

1.3 Teaching Ideas and Talent Training Ideas

In the first place, we adhere to the teaching concepts of "innovative education" and "student-oriented". We regard education as the foundation of professional construction, quality as the guarantee of professional survival, innovation as the driving force of professional development, and characteristics as the cornerstone of professional development. Emphasis are placed on improving students' employability, entrepreneurship, international communication abilities, and lifelong learning abilities.

In personnel training, we seize the opportunity of the MoE Outstanding Engineer Project, follow the law of education and teaching, and in accordance with the needs of the development of socialist market economy, adhere to the principle of "Solid foundation, emphasizing practice, strong ability and innovation". We attach importance to all-round quality education, innovation and entrepreneurship, and build a high-quality, compound and internationalized talent training model.

1.4 Reform in Education

In terms of teaching reform, we adhere to student-oriented initiative, deepen the reform of teaching content and curriculum system, integrate teaching content with the international development trend of disciplines, promote the innovation of curriculum and practical teaching system, deepen the reform of teaching methods, actively implement heuristic and discussion-based teaching, highlight the cultivation of innovative ability and the development of students' personality, and vigorously promote bilingual teaching. Modernization and internationalization of teaching methods will promote the application and popularization of international educational technology such as Internet and information technology; updating and optimizing teaching contents so that students can familiarize themselves with the current situation and trend of development in civil, material and related industries at home and abroad, master basic knowledge and professional knowledge and skills in the field of civil and material science, get engaged in civil engineering, material science and engineering, etc, develop the ability of related field. Thus we need to optimize and innovate teaching system, establish and improve teaching operation mechanism, management system and teaching evaluation system, and make it meet the needs of international talent training.

2 Internationalization Model of Talent Training

2.1 The Integrative Teaching Model: in-and-off class; in-and-off school; at home and abroad

Professors are often invited to lecture or teach for some time on a regular basis. To name a few: academicians of the Russian National Academy of Sciences, Professor Folker. H. Wittmann of the Federal University of Technology of Switzerland (Zurich), Professor Gilian Gerke, Professor Harald S. Mueller of the Karlsruhe Institute of Technology of Germany (formerly the University of Karlsruhe), Professor Rouss, Vice-President of the State Krastar University of Technology, Professor Li Zongjin of the Hong Kong University of Science and Technology and Dr. Liu Min of the North Carolina State University. At the same time, students are organized on a regular basis to visit Germany (once every two years), the United States (once a year), Australia, China and other countries and regions for short-term exchanges; On this basis, from 2018, we jointly launched and hosted the first Sino-German Student Conference with the Magdeburg University of Applied Technology in Germany; The First International Students Conference on Civil Engineering, Environmental Engineering and Material Science (CEEMS). More than 150 students from China, Germany and Belarus gathered to discuss the related contents in the fields of civil engineering, environmental science and engineering, and material science and engineering. Through the above measures, we can form an integrative teaching mode of talent cultivation in and out of class, in and out of school, at home and abroad. Integrating international talent training with cooperation in running schools at home and abroad, co-sponsoring international conferences and jointly conducting scientific research work, thus forming a long-term and stable platform for international cooperation, providing students with a platform for understanding the world, and for communicating and exchanging with them on an equal and friendly basis. All this has enabled students to have a broader international perspective, gradually integrate with the world, and enhance international competitiveness.

2.2 Sino-US Joint Program in Civil Engineering

In 2016, our college and the Civil Engineering College of Kansas University jointly applied to the Ministry of Education for the Sino-US Joint Program in Civil Engineering and was approved. The project aims to cultivate advanced applied talents with innovative thinking and international vision, strong engineering practice ability and innovative ability. The mode of running the program is "3+2". After completing the first three years of study at QUT, students can choose to go to Kansas University for two years to further the study. Talent training program, teaching plan and the whole teaching process are formulated and implemented by the experts and teachers of the two schools. The curriculum system draws on the advantages of the two schools, maintains Chinese characteristics and integrates into the United States at the same time, and is in line with international standards. The Sino-US Joint Program in Civil Engineering has promoted the internationalization of talent training in our college. It has further expanded the platform of cooperation and exchanges and has laid a good foundation for the training of internationalized talents in our college.

2.3 The Training Model of Dual-tutor System

Talents are trained by tutors from both Schools and Enterprises. Engineers with foreign working experience are invited and employed as enterprise tutors, who work jointly and closely with in-school tutors. This mode plays an effective role in enriching foreign engineering experience. As the enterprise engineering tutor group is composed of project managers and engineers, it makes the professional teaching more integrated, refined and internationalized through the mutual exchange and supervision between the two different tutor groups.

2.4 The Internationalization Model based on "Overseas Project (Group) Learning" and Professional Information

Based on the "Overseas Project (Group) Learning", students are familiar with the basic methods of production, construction and maintenance of engineering materials at different stages of practice in schools and enterprises, thus able to improve their comprehensive application of knowledge, understand the actual needs of engineering, and cultivate their comprehensive engineering abilities such as professionalism, analysis, communication, solidarity and cooperation, management and expression. Therefore, they develop the ability to independently engage in material design and operation, analysis and integration, research and development, management and decision-making in a certain direction in the field of material science and engineering. In this way, they will meet the needs of material science and technology development and social progress in the future, and become high-quality talents with practical ability, innovation ability and international vision for the future.

The mode of internationalization of professional information is to promote students' professional learning of the forefront of international disciplines, and to work with the best teachers and students, to be conscious of the most prominent hot issues and the most advanced professional technology and make them aware of the development of the times, the country and society, to be closer to the frontier of disciplines, to have more professional and advanced technical means, and thus to be able to closely integrated with the requirements of the times, society and the country. It is an effective and efficient way of training students' engineering innovation ability.

3 The Internationalization of Innovative Curriculum System

By fully investigating the current situation of relevant universities and majors at home and abroad, by analyzing the demand for internationalized talents and by combining the characteristics of our university's personnel training code of "consolidating professional foundation, strengthening practical ability, highlighting innovative spirit and accentuating internationalization consciousness", our reform curriculum system fully highlights engineering practice education, innovative research and internationalization education, and consolidates the foundation of "thick foundation" general education curriculum construction. We pay attention to the integration of the basic courses of various specialties, further strengthen the construction of the platform of the series of specialty courses, and pay attention to the coherence, continuity, frontier and internationalization of the courses. At the same time, we try to introduce the group system of internationally advanced courses (video open courses) related to disciplines, so as to make the new group system more suitable for the training of innovative applied talents in civil engineering, material science and engineering under the new situation. Relying on the international team of teachers and the national bilingual excellent courses, we offer professionally cutting-edge lectures and bilingual courses to our students. We also highlight the international characteristics of the frontier disciplines, cultivate students' international vision, and integrate the international vision with the foreign language communication skills.

4 Internationalized Faculty Team

Focusing on improving the quality of faculty in an all-round way, we try to improve the academic level, teaching and scientific research ability and international cooperation and communication ability of young and middle-aged backbone teachers, and cultivate an international team of teachers with international vision and international cooperation and communication ability.

4.1 Focus on the Training of Young Teachers: Through the formulation and implementation of a series of plans and systems, such as the Young Teachers Navigation Plan and the Management Measures for Young Teachers' Domestic (Foreign) On-post Training

Plan, we deepen the reform and innovation of young teachers and promote the structural optimization and rational allocation of teachers' resources. We continue to improve the quality of talent introduction, retention and utilization in line with international standards, focus on the training of subject leaders and key teachers with a good working mechanism. At the same time, efforts are made to create an excellent international working environment suitable for young teachers' growth, make good use of existing talents, train key talents, introduce urgently needed talents, reserve future talents, and strive to build a professional-part-time, specialized-oriented team of international teachers.

4.2 Make full use of international and domestic high-quality educational resources, speed up exchanges and cooperation with world-renowned universities, meet the needs of international running schools, and promote overseas training programs. Taking the National Fund for Overseas Studies and the Shandong Fund for Overseas Studies as opportunities, 2-4 young teachers are sent to famous foreign universities and key domestic universities every year as visiting scholars, and a number of high-ranking teachers are sent to conduct academic exchanges and visits abroad. Academicians and renowned scholars at home and abroad are invited to make academic reports and team building for teachers and students majoring in material science and engineering. This broadens the teachers' international horizons and international cooperation and exchange capabilities, and improves their level in teaching.

4.3 With the help of Marine Environment Concrete Technology Innovation Base, we have recruited professors from Germany, Britain, China and other countries and regions. The academic master leading the overseas teams is Professor Wittman, an internationally renowned expert in the field of civil engineering, material science and engineering, and academician of the Russian National Academy of Engineering. Facing the "Belt and Road" initiative, we focus on the demand in the construction of long-life concrete materials and structures in marine environment and focus on the basic theory of durability of marine structures, new concrete structure system, marine corrosion and protection, which are closely related to major projects[1-3].

4.4 The "Belt and Road" national initiative has been integrated into the Asia Pacific region, the Western European continent and the Central South Asian countries. On the basis of strengthening the existing international cooperation and exchanges with universities such as University of Kansas, Germany, University of Karlsruhe, Magdeburg Applied Technology University in Germany, TU

Clausthal in Germany and Australia University in West Asia, we have actively explored new cooperation opportunities of the countries along the route, such as the Belarusian National Technical University; Azerbaijan Technical University, Azerbaijan University of Architecture And Construction. These universities are deeply integrated with the "Belt and Road" initiative and contribute to the widening and perfecting of the platform for international cooperation and exchange.

After several years of efforts, we have built up an international teaching staff including "Russian Academician", "Famous Foreign Experts and Scholars", "Humboldt Scholars", "Senior Visiting Scholars" and "Returned Overseas Students". We have accumulated abundant international cooperation resources and relevant experience, which has laid a solid foundation for promoting China's role in building a better world and cultivating more professional talents[4-6].

5 Conclusions

5.1 Combined with our own school-running characteristics and academic resources, we highlight key points and characteristics, actively participate in the "Belt and Road" initiative and establish an international education system, including the overall goal of specialty construction, teaching ideas and personnel training ideas and teaching reform.

5.2 We explore and carry out the reform and practice of international talent training mode and innovative curriculum system, and form the integrated teaching mode of integrating teaching in class and off class, in school and out of school, at home and abroad. Besides, we establish Sino-American cooperation program, the school-enterprise dual-tutorial training mode, the learning mode based on "Overseas Project (Group) Leaning" and the internationalization mode of professional information, etc.

5.3 We attach importance to the internationalization of the teaching staff, pay attention to the training of young teachers, make full use of international and domestic high-quality educational resources and "111" Intellectual Base, broaden and improve the platform for international cooperation and exchanges, and build an international team of faculty.

References

[1] Masalimova A R, Shaidullina A R. Study of International Mentoring and Coaching Practices and Their Constructive Application in the Russian System of Corporate Education and Training[J]. International Journal of Environmental and Science Education, 2016, 11(8): 1797-1806.

[2] Alsharari N M. Internationalization of the Higher Education System: an Interpretive Analysis [J]. Interna-

tional Journal of Educational Management, 2018, 32(3):359-381.

[3] Lyu Ping, Lu Guixia. Teaching Content Optimization and Teaching Mode Reform and Practice of "Green Building and Technology" Course [J]. Research on Higher Engineering Education, 2018, 58-60.

[4] Kirk S H, Newstead C, Gann R, et al. Empowerment and Ownership in Effective Internationalization of the Higher Education Curriculum [J]. Higher Education, 2018, 76(6): 989-1005.

[5] Diao P H, Shih N J. Trends and Research Issues of Augmented Reality Studies in Architectural and Civil Engineering Education—A Review of Academic Journal Publications [J]. Applied Sciences, 2019, 9(9): 1840.

[6] Sun X, Jia Y, Li Z, et al. Comparison of Engineering Education in Norway and China [J]. International Journal of Higher Education, 2018, 7(1): 98-102.

(3) From Students, for Students, Let's Build the Future

Liu Junwei, Miao Jijun, Gilian Gerke, Lyu Ping, Cui Yifei, Liu Yanchun,
Liang Zuodong

School of Engineering Qingdao University of Technology

Abstract

The first QUT-HSMD International Student Conference is an event hosted by Die Hochschule Magdeburg-Stendal (FH) (HSMD) and Qingdao University of Technology (QUT) which bring together Chinese and German students, as well as academic organizations, to learn, exchange ideas, innovate and collaborate to support undergraduates and postgraduates in their personal, professional and academic development.

Key Words

QUT-HSMD, International Student Conference

The first QUT-HSMD International Student Conference was successfully held in the Academic Lecture Hall of the North Campus of Qingdao University of Technology (QUT) from October 12 to 14. More than 250 people including the student delegates from the Magdeburg University of Applied Sciences and the Belarusian University of Technology, as well as the Chinese student representatives, attended the academic conference. The conference was hosted by Junwei Liu, who is the Vice Dean of the School of Civil Engineering. This academic conference was co-hosted by Qingdao Technological University and Magdeburg University of Applied Sciences, Germany, and undertaken by the School of Civil Engineering. It aimed to deepen exchanges and cooperation between students in the fields of civil engineering, environmental engineering, and green technology, enhancing collaboration and communication in terms of researching, teaching, training and etc.

The opening ceremony of the QUT-HSMD International Student Conference was held in the Academic Lecture Hall of QUT, Shibei Campus on October 12th. Professors Gerke and Monsees of Magdeburg University of Applied Sciences in Germany, Professor Алексей of the Belarusian University of Technology, Vice President Dehu Yu of QUT, Director of International Exchange Office Chuntang Liu and leaders of the School of Civil Engineering attended the opening ceremony. First of all, the Vice-President of QUT Mr. Dehu Yu gave a welcoming speech to German and Belorussian students, believing that via this academic conference, students from various schools could strengthen the connection, share scientific research results, and inspire scientific research ideas; understand the academic frontiers in the field of civil engineering and similar fields, and understand industry trends. After that, Jijun Miao, Dean of the School of Civil Engineering, Professor Lyu Ping, Head of the Chinese Delegation, and Professor Gerke, Head of the German Delegation, delivered speeches successively, introducing the history of QUT and the profound friendship with Magdeburg University of Applied Sciences over the years[1].

A total of 30 reports were organized at this conference, which fully demonstrated the research results of the two universities in civil engineering, environmental protection, and new materials. The atmosphere of the conference was enthusiastic with active question-and-answer sessions. On the afternoon of October 14th, Mr Rainer Fnitsche, Deputy Mayor of Magdeburg, Germany, attended the discussion and delivered a speech accompanied by Vice President Dehu Yu. More specifically, Mr Rainer Fnitsche introduced the history and development of Magdeburg, compared and analyzed the differences between Magdeburg and Qingdao, and also expressed his affirmation and appreciation for the convening of the first International Student Conference.

After two days of exchanges and contacts, the student representatives of both sides begun to be familiar. All students actively interacted, producing a strong academic atmosphere surrounded the venue. The students of QUT who participated in the conference actively asked questions, which demonstrated the spirit of active exploration, continuous improvement, and active study of science. Meanwhile, It also demonstrates the university's spirit of learning and the students' qualities of seeking truth, being pragmatic, rigorous and diligent. Overall, the presentation schedule of the QUT-HSMD conference is listed below.

Study Presentation Schedule			
13 Oct			
Time	Topic	Name	Host
9:00	Microencapsulated Self-Healing Material	Zhang Yingrui	Zhang Yingrui
9:20	Structures Made of Waterproof Concrete	Franziska Kristin Radtke	
9:40	Research on Viscoelastic Damping Materials and Damping Structures	Wu Di	
10:00	Surprises During Remediations	Laura Pelzer	
10:20-10:35 Coffee Break			
10:35	Steel Beam-to-Column Connection Panel Zone with Unequal Beam Depths	Zhao Fei	
10:55	Rewetting of Fens: A Case Study from Northern Germany	Florian Schlomo Hetzel	
11:15	Water Capillary Absorption of Alkali Slag Concrete After Salt Frost Action	Han Xukang	
11:35	Formation and Succession of Oxbow Lakes in the Middle Elbe Biosphere Reserve	David Joël Hunger	
11:55	New Materials for the Treatment of Printing and Dyeing Wastewater	Zhang Yue	

(3) From Students, for Students, Let's Build the Future

(续表)

12:15-13:00 Lunch		
13:30	Decentralised Rainwater Management	Plett Viktor
13:50	Carbon Source Supplement Strategies in Black Water Treatment System	Li Siyu
14:10	Tunnel Construction	Isabell Göhre
14:30	Research on the Vibration of Subway with Disc-Spring and Thick Rubber Isolator	Zhao Hanzhu
14:50-15:05 Coffee Break		
15:05	Supporting An Excavation Pit With Steel Piling	Laura Niens
15:25	Study on Seismic Behavior of Steel Structure Joints with Inner Diaphragms and Outer Ring Plates	Ju Jiachang
15:35	Construction and Design of Skyscrapers	Jannik Lukas Wiechens
15:55	The Cultivation of College Students' International Vision	Zheng Haonan
16:15	Flood Protection Measures	Mark-Philipp Weber
16:35	Briefly.... Campus life	Cui Hanwen
14 Oct		
9:00	TMD for Swing Vibration Control of Suspended Structures	Wang Hao
9:20	How to Make A High-Speed Railway Tunnel Safe For A Case of Emergency	Manuel Kempf
9:40	The Mechanical Mechanism of Pipe Pile	Zhu Na
10:00	Fire Protection Engineering - Preventive Structural Fire Protection and Engineering Methods in Germany	Pascal Leonard Haak

(The "Li Xi" label spans the 13:30–16:35 block on 13 Oct.)

(续表)

14 Oct		
10:20-10:35 Coffee Break		
10:35	Synthesis and Properties of Flame Retardant Polyurea Elastomers	Wang Rongzhen
10:55	Hoover Dam - a technical and engineering wonder"	Mareike Ines Rathge
11:15	Molecular Dynamics Study on the Adsorption Properties of Ions on the Surface of Sodium Alumino - Silicate Hydrate(NASH) Gel	Zhang Jinglin
11:35	Illegal Dumping - An Environmental Challenge	Franziska Christa Söhnholz
11:55	The Characteristics of Underground Comprehensive Pipe Gallery Structure	Han Jinpeng
12:15-13:00 Lunch		
13:30	Electronic Waste - Challenges and Chances	Sven Hartmut Mauch
13:50	An Efficient Repair Mortar	Lu Yu
14:10	The Elution of Heavy Metals at Tailings Piles	Sven-Simon Lattek
14:30	Influence of Reinforcing Ring Parameters on Seismic Performance of Steel Structure Joints	Zhang Zhipeng
14:50-15:05 Coffee Break		
15:05	The Elimination of Phosphorus Sewage Treatment Plans	Vincent Weber
15:25	Preparation of Magnetic Catalyst	Yang Yuying
15:45	Long-term Study for the Waste Behaviour of Different Population Groups in Germany	Jonas Thiel
16:05	Chinese Communication	Hu Yuan
16:25	Urban Mining: Discover the Value of Building Material Waste	Julia Marie Zigann
16:45	The Culture of Chinese Festival	Ren Zhongyuan

(The "Zhu Na" label spans 10:35–11:55; "Wang Hao" spans 15:05–16:45.)

Reference

[1] Lyu Ping, Lu Guixia. Teaching content optimization and teaching mode reform and practice of "green building and technology" course [J]. Research on higher engineering education, 2018, 58-60.

(4)Research on the Training Mode of Applied Innovation Talents in the Major of Material Science Engineering-take Qingdao University of Technology as an Example

Lyu Ping, Wan Fei, Che Kaiyuan, Feng Chao, Ma Mingliang, Liang Longqiang

Research Institution of Functional Materials Qingdao University of technology

Abstract

In response to the new strategic needs of material-based application-oriented innovative talents in the current countries and regions, the new talent training mode and the corresponding curriculum system will implement profound reforms with the times according to the characteristics and advantages of this specialty. Talent cultivation, talent training mode innovation, reinforcement of innovation and entrepreneurship ability as well as training of international competitiveness will be realized via the subject platform and the practical teaching platform while optimize the application-oriented innovative talent training mode of materials science engineering under the new situation.

Key Words

Applied Innovative Talents; International Competitiveness; Talent Cultivation Model; New Entrepreneurial Ability

At present, several major strategic decisions such as Enhancement of Independent Innovation Capability, the Belt and Road and National Ocean Power Strategy have been put forward. In the Twelfth Five-Year Plan for Medium and Long-Term Science and Technology Development Plan, the material science will be applied as a forward-looking deployment basic research and frontier technology research field, which proposes to focus on the development of green building materials, renewable energy materials and their application technologies integrated with buildings. In the "12th Five-Year Plan for Green Building Science and Technology Development", the development of high-service characteristics and long-life green civil engineering materials will also be the key technical issues. The national "Made in China 2025" requires vigorous development of new materials such as functional polymer materials and special inorganic non-metallic materials to promote the development of green materials technology. All of this means that the country is facing a major demand for innovative materials professionals.

Qingdao has been given the status of a leading city in the core area in the implementation of the National Ocean Power Strategy as well as the construction process of the Shandong Peninsula Blue Economic Zone and the Qingdao West Coast New District National Innovation Zone. To realize the rapid development of the marine industry and take development as a support, application-oriented innovative talents who have a solid foundation of natural sciences, a foundation of humanities and social sciences and a foundation of science and engineering, with strong engineering practice ability, self-acquisition ability, innovation and entrepreneurship ability and international competitiveness, need to be cultivated in materials science and engineering.

In view of the new strategic needs of materials and application-oriented innovative talents in the current country and region, new positioning and target optimization for material science and engineering professionals training are required according to the characteristics and advantages of this specialty, and subjects are constructed according to local conditions. On this basis, we will carry out profound reforms of the new talent training model with the times, via strengthening the cultivation of students' innovative and entrepreneurial ability, engineering practice ability and international competitiveness, to meet the current social development needs of the country and the region better.

1 Achieve Telant Training VIA the Subject Platform and Practical Teacgubg Platform

Eleven innovation platforms are taken as national and provincial teaching resources including The National Institute of Higher Education Discipline Innovation and Innovation Base (National Environmental Concrete Technology Innovation and Intelligence Base), National Experimental Teaching Demonstration Center (Qingdao University of Technology Civil Engineering Experimental Teaching Center), Marine Environment Concrete Technology Ministry of Education Engineering Technology Center, Blue Construction and Safety of the Economic Zone, Shandong Province Collaborative Innovation Center, Shandong Provincial Key Laboratory of Concrete, and Shandong Province Concrete Structure Durability Engineering Technology Research Center. Most teachers in materials science and engineering have undertaken a number of national-level projects, and the research topics are closely integrated with major projects such as the Hong Kong-Zhuhai-Macao Bridge and the Qingdao Metro. We will create a driving mechanism for the cultivation of engineering innovation capabilities via these scientific research projects, thus comprehensively develop students' innovative research capabilities in a three-dimensional manner. At present, the number of research projects that students participate in is a total of 23, with an open talent training model. The latest achievements and concepts are transmitted to the students so that students can keep abreast of the cutting edge of research. At the same time, students are allowed to participate in the research of teachers, especially major engineering projects, taking practical teaching platform and innovation base as the innovation and entrepreneurship chain, which effectively strengthen the cultivation of students' ability of innovation and entrepreneurship. Based on the Materials Science and Engineering Practice Education Center and the practice bases, all-round cooperations between schools and enterprises are carried out while four bases including student practice base, student employment base, teacher training base as well as technology research and development base are constructed. On the basis of co-constructing enterprises to deeply participate in the training and solving the problem of practical training of talent cultivation, company selects and hires employees to study at the college regularly, while college employs high-level engineering experts with rich experience to teach or part-time. The "dual-master" teachers are trained via the practical teaching platform, and teams of "dual-master" echelon-based teaching teams with reasonable school-enterprise cooperation are formed to carry out talent training in a three-dimensional manner.

2 Innovation of Talent Training Model

According to the needs of enterprises, it is necessary to formulate talent training programs to develop innovative practical talents. Rich practical experiences of the engineering instructor group of the institute and the full-time teachers of school are effectively exerted via talent training through the "dual-master" of school-enterprise cooperation. Through the mutual exchange and supervision between these two different tutor groups, the undergraduate teaching of materials science and engineering is more integrated and engineered.

The formation of a reasonable school-enterprise "dual-master" echelon-style teaching team, in which the team of full-time teachers is composed of full-time teachers with engineering experience, and the person in charge with rich engineering experience and leadership, takes the position as the team leader; experts with engineering experience serve as consultants. Meanwhile, the team of off-campus part-time teachers is composed of industry-related enterprise engineers; industry-related enterprises and business leaders take the position as the team leaders, and influential experts in the industry hold a post of expert part-time teachers. In the process of teaching, "the three-stage, three-theme" talent training mode and the basic construction of the school classroom are carried out, while basic knowledge theory and basic literacy training of students are solved through the courses of materials science and materials science research methods. The innovative ability and engineering practice ability of students have been cultivated through the establishment of professional front-end courses by team of full-time teachers, and have further developed through the enterprise new material preparation process and new technology research by the off-campus part-time teachers.

The ability to comprehensively apply knowledge of students has been effectively improved through internships at different stages of school and business, which can be familiar with the basic methods of production, construction and maintenance of engineering materials. Via understanding the actual needs of the project, the ability of engineering, such as professionalism, analytical ability, communication and communication ability, solidarity and cooperation ability, management ability and expression ability has been cultivated to meet the needs of future material science and technology development and social progress, which gets the ability to independently design and operate materials, analyze and integrate materials, research and development, management and decision-making in a certain direction in the field of materials science and engineering. Thus, students will become future-oriented high-quality talents with innovative entrepreneurial ability and international competitiveness[1-9].

3 Strengthen the Cultivation of Innovation and Entreneurship

3.1 Professional Competition Mode to Cultivate Innovative Ability in the Way of Competition

Through the diversified information brought by the competition, a three-dimensional, long-term and open-ended talent training system with professional competition and international curriculum as a communication and cooperation chain will be built, while professional vision of students will be expanded. Thus the enthusiasm of students for learning will be enhanced, innovative and entrepreneurial ability as well as international competitiveness will be cultivated. Through the open-ended talent training system, various competitions organized by domestic professional organizations are included in the teaching system as a topic, at the same time a series of competitions such as college students' science and technology innovation contests and business plan competitions are actively carried out in the period of vacation.

The extracurricular practice activities build a training platform for students, taking various competition activities at the national, provincial and school levels as incentive points, opening the laboratory as an innovative operation mechanism. In conjunction with the relevant regulations of the credit system training program, students can obtain innovative credits by applying for university students' innovation and entrepreneurship projects, participating in science and technology competitions, and teacher research projects. The school selection competition such as a school-level high-strength concrete design competition, pumping concrete design competition will be set up every year, and winners will be recommended to participate in provincial and national science and technology competitions. Students are encouraged to participate in the forum for academic reports to enhance the enthusiasm for learning and promote the improvement of all-round capabilities.

3.2 Incorporate Innovation and Entrepreneurship Training into Talent Training Programs

According to the new talent training objectives and the innovative curriculum system and the reform policy of the talent training program, further requirements of students for innovation and entrepreneurship are stipulated. In the training program, students are required to develop and innovate, including strong sense of innovation and initial ability to innovate; students are required to master the basic methods of scientific thinking and scientific research, so that they can get the basic ability to independently acquire knowledge, ask questions, comprehensively analyze problems and solve problems, and have strong engineering awareness, value-benefit awareness and innovative spirit. In the curriculum system compulsory course module, the innovation and entrepreneurship course group is set up. In the talent training program, the requirement of minimum innovation of 8 credits for innovation and entrepreneurship is also put forward. The learning of innovation and entrepreneurship practice runs through the entire learning life of the students in the 1st to 8th semester[10,11].

Students can earn innovative and entrepreneurial credits in the following ways: ① Participate in special competitions such as technological innovation, creative design, and entrepreneurial programs; ② Participate in the Innovation Entrepreneurship Training Program; ③ Obtain relevant professional qualification certificates in the field of materials science and engineering; ④ Participate in school and enterprise scientific research and technical research; ⑤ Participate in more than 10 academic reports related to this major and submit a summary report; ⑥ Study abroad or conduct joint training outside the country; ⑦ Obtain authorized national invention patent; ⑧ Publish core journals or SCI/EI research papers.

4 Focus on the Cultivation of International Competitiveness

When "the Belt and Road" major strategy is continuously promoted and the construction of the blue economic zone is being implemented in Shandong, the Materials Science and Engineering Department of Qingdao University of Technology focuses on the economic development of the country and the region, emphasizing a more three-dimensional, long-acting and open-ended talent training system, focusing on cultivating talents with innovative entrepreneurship and international competitiveness. To achieve this talent training goal, the mode in the process of student development is reformed mainly through the following three aspects[12].

4.1 Improve the Construction of Internationalized Characteristic Curriculum Groups

Improve the professional foreign language literacy of students by setting civil engineering materials (bilingual), concrete (bilingual), and materials science (bilingual) in the curriculum system. Through the advanced courses of advanced civil engineering materials and marine coating materials series, the professional information internationalization model is used to push students' professional learning to the forefront of international disciplines, and the best teachers and students, the most prominent hot issues, and the most advanced professional technology. For the sake of the times, it will pay more attention to the development of the times, the country and the society, and be closer to the frontiers of the disciplines, with more professional and advanced technical mean.

4.2 Promote the Internationalization of Academic Exchanges Actively

Promote cooperation and exchanges with universities and organize academic conferences at home and abroad actively. Through the introduction of special reports, lectures and forums by experts and scholars, including academicians, students can learn the cutting-edge knowledge of international disciplines. Via carrying out international exchange activities among students actively, applying for the National Scholarship Committee "Excellent Undergraduate International Exchange Program" and other projects to support students to study abroad, the understanding of professional frontier knowledge and professional development trends have been effectively improved.

4.3 Promote International Talent Training via International Scientific Research Projects and Scientific Research Platforms

The international natural science research projects of the National Natural Science Foundation will be integrated into the international talent training process, enabling students to learn the cutting-edge expertise through international research projects, which effectively expand their international vision. Some initiatives including introducing famous overseas experts and scholars, inviting overseas academic masters and overseas academic backbone to cooperate and exchange, sending young teachers to study at the prestigious schools where overseas experts are located, as well as jointly holding international academic conferences in this field, are implemented via utilizing the national "111 Plan" Qingdao University of Technology, marine environmental concrete technology and other innovative intelligence base platforms, which promote the cultivation of international talents.

5 Conclusions

According to the goal of this professional talent training model and the teaching characteristics of this major, we have actively carried out the reform of several training models such as school-enterprise cooperation training, international joint training, excellent engineer training plan, entrepreneurship training, innovation training, etc., which provides institutional support for the individualized development of students and formulating a scientific and rational talent training program. The innovation of the talent training model promotes the cultivation and improvement of students' abilities.

6 References

[1] Shi Hui. Research on the training Mode of innovative talents in Colleges and Universities [D]. Tianjin University, 2011.

[2] Chen Zhigang, Yang Xinhai, Wu Jianrong, Fu Baochuan, Xu Zongning, Zhang Xiwu. A study on the training Mode of Engineering innovative talents in Local Colleges and Universities-A case study of Suzhou Institute of Science and Technology [J]. Research on higher Engineering Education, 2012, (01):75-80.

[3] Zhang Xinyue, Dong Shihong, Zhou Jinqi, Wei Ling. Thoughts on the cultivation Model of Applied innovative talents in undergraduate course [J]. Research on Educational Development, 2008 (Z1):122-124.

[4] Lu Shiyan, Guo Dehong. Discussion on the Reform of the training Mode of Applied innovative talents [J]. China University Science and Technology, 2015 (04):87-90.

[5] Jin Baohua, Liu Yu Ham. The cultivation Mode and Enlightenment of Applied and innovative talents in developed countries [J]. Teaching and Research, 38 (01):87-92.

[6] Dong Xiaomei. Discussion on the training Mode of internationalized and innovative talents in Local Engineering Colleges [J]. Innovation and Entrepreneurship Education, 2012, 3(02):72-75.

[7] Wang Xiaolai, Li Zhixia, Chai Xin, Yang Heheng, Wu Yuanxing. Research on the achievements of "Industry, Learning, Research and Innovation" under the Mode of cultivating Applied innovative talents[J]. Science, Technology and Innovation, 2018 (06):116-119.

[8] Cai Xinhai. A study on the Reform of the training Mode of Engineering undergraduate talents in China [D]. South China University of Technology.

[9] Zhang Weijun, Bai Shuxin, Wu Wenjian, Jiang Dazhi. Study on the Reform of training Program of Materials Science and Engineering Talent [J]. Journal of higher Education Studies, 2011, 34 (03):18-20.

[10] Wang Zhangzhong, Pi Jinhong, Ba Zhixin. Thoughts on the cultivation of Applied talents in Materials Science and Engineering Specialty [J]. Journal of Nanjing Institute of Engineering (Social Sciences Edition) 2007 (01):37-40.

[11] Yang Yuan, Xiao Guoqing, Xu Delong. Innovation and practice of the

training Mode of material Science and Engineering Specialty [J]. Journal of Xi'an University of Architectural Science and Technology (Social Sciences Edition), 2015 (01): 97-100.

[12] Zhang Xiaoyan, Xiang Song, Li Yuanhui, Lei Yuanyuan, Li Wei, Wan Mingpan. Construction of Professional Curriculum System under the Mode of Cultivating Innovative Talents in Materials Science and Engineering[J]. Education and Teaching Forum, 2013 (41): 216-217.

2. Civil Engineering

(1)Fire Protection Engineering Preventive Structural Fire Protection and Engineering Methods in Germany

B. Sc. Pascal Haak

Department of Water, Environment, Construction and Safety

University of Applied Sciences Magdeburg

Abstract

Both in new construction projects as well as in changes of use, the safety regulations must always be complied with. These regulations limit the possibilities of scopes for design and options for use. Deviations are possible, even then protection goals must be adhered strictly. Engineering methods are suitable as verification for compliance with these protection goals for structural deviations. This paper briefly introduces common fire protection engineering methods and presents computer-based fire simulations and evacuation simulations. Finally, the advantages and disadvantages are briefly explained.

Key Words

Fire Protection, Fire Siluation, Engineering Fire Protection.

1 Introduction

A variety of different building regulations make demands on fire protection.

According to four protection goals are defined[1], prevention of outbreak of fire, prevention of fire and smoke spread, enabling of rescue of humans and animals, and enabling of effective extinguishing works.

The design and usage possibilities of a building are limited by the standardized regulations. Deviations from the rules are possible but must be considered. It must be proven that the requirements belonging to the protection goals are met. Proof of compliance with the protection goals and the achievement of an equivalent level of safety can be provided by engineering methods.

2 Engineering Methods in Fire Protection

Engineering methods offer many solutions for fire protection issues. Possible applications are for example evidence of fire effects, smoke spreading and smoke exhaust, or personal safety and evacuation.

According to the application of engineering methods is divided into the following sequence [2]:

(1)qualitative analysis

(2)quantitative analysis

(3)comparison between result and requirement

(4) presentation and interpretation of the results

Different calculation methods can be used for the quantitative analysis.

The possibilities range from simple manual arithmetics to multi physic calculation models. The simulations can better respond to individual building characteristics and provide more accurate results for a specific scenario. At the same time however, they are associated with more effort and experience.

In extracts, two engineering methods are presented:
• Fire and smoke simulations
• Pedestrian flow analysis (simulation)

2.1 Fire and Smoke Simulations

The spread of fire and smoke can be simulated with computers by using CFD (computational fluid dynamics) models. The first step is to define a fire scenario. Here, the location of the fire and the burning material are determined. In the second step, the parameters for a design fire are defined. Instead of the standard temperature-time curve of the ISO 834, which is used for fire resistance testing of components, the design fire is usually modelled by a so-called natural fire. The course of such a design fire is shown in Figure 1.

Parameters include rate of fire spread, maximum heat release rate, fire load, combustion efficiency and product yields.

In order to be on the safe side with the results, the fire scenario should be a worst credible scenario and the parameters of the design fire are chosen conservatively.

The results are compared with the requirements such as the layer height of smoke or a maximum temperature at a defined point.

In the final interpretation, a statement can be made as to whether the functional requirement of a protection goal is met or not.

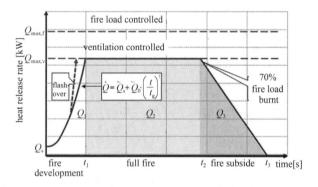

Figure 1 Design Fire According to [2]

2.2 Pedestrian Flow Analysis (escape simulation)

With high numbers of people, the question often arises how long an evacuation takes. The total evacuation time consists of individual time periods: the detection time, the alarm time, the reaction time and the escape time. The timing is illustrated in Figure 2. Computer simulations can be performed to determine the evacuation time. They use microscopic models, in which each person is contemplated individually[2,3]. By considering individual persons, individual characteristics such as the reaction time and gait velocity are assigned. In addition, in the model the building characteristics such as the escape route guidance and width can be accurately represented. This provides a microscopic model with the most accurate results. The calculated evacuation time is compared with the available time. Safety is achieved if the evacuation time is less than the time available. This always takes into account a safety factor.

The available time can be determined, for example, by a fire simulation in which the radiation and convective heat of the fire or smoke are proven for a sufficiently long time.

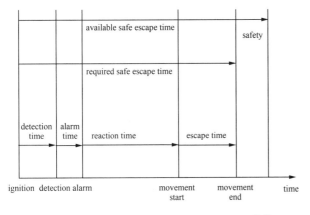

Figure 2 Escape Time Expiration According to [2,3]

3 Conclusions

Engineering methods provide solutions to many fire protection issues. In particular, the simulations can respond to a precise scenario or a specific building and provide an individual solution. This allows complex buildings and safety requirements to be reconciled. However, care must be taken in the choice of input parameters. In addition, the simulations, unlike the simple calculation methods, require a high computational cost. Since the handling of the simulation programs is very complex and the possibilities are not known to many people so far, they are little used. Due to the possibilities they offer and the ever-improving computing performance, they will become more important in the future.

References

[1] Musterbauordnung in der Fassung November 2002, zuletzt geändert am 13.05.2016.

[2] Technischer Bericht vfdb TB 04-01vfdb Leitfaden Ingenieurmethoden des Brandschutzes, Braunschweig, 2013.

[3] Richtlinie für mikroskopische Entfluchtungsanalysen, Version 3.0.0, 10 März 2016, © 2004-2016 Rimea e. V.

(2)Construction and Design of Skyscrapers

Jannik Lukas Wiechens

Department of Civil Engineering, University of Magdeburg-Stendal

Abstract

Throughout the history, humans have attempted various ways of building highrises. The first skyscrapers began to appear in the cities in the late 19 century as a result of technological breakthroughs in building materials and methods, including reinforced concrete, steel, and elevators. Buildings of this kind bear three types of loads: dead loads, live loads and environmental loads. All loads are resolved into vertical and horizontal forces on the structure. Typical structural systems used in skyscrapers include core and outrigger structures, steel frames and trusses.

Key Words

Skyscrapers, Environmental Loads, Structural System

1 Introduction

The urge of people to strive for height is by no means a development of modern times. In the past and in antiquity, there are several examples of tall buildings that have emerged from different intentions. Structures of the past such as the Lighthouse of Alexandria, the pyramids of the ancient Egyptians or the Tower of Babel illustrate that even the humankind tried to build tall and visible structures.

The human has always had the impulse to push the limits and overcome boundaries. The incentive for further development, architecture and improvement of the technical conditions, are still in evidence. These points are reflected to a high degree in today's modern high-rise buildings and show that conduction is different in every area of the world.

Nevertheless, the actual skyscraper was formed from the requirements and conditions of its own epoch and was the very expression of American architecture[1].

Thus, the world's first skyscraper was built in 1885 in Chicago by a civil engineer William Le Baron Jenney. It was called the "Home Insurance Building". It had a height of 42 meters and ten floors. The city is located in the north-eastern part of the United States and is therefore referred as the "cradle of high-rise culture". Since then, the height of skyscrapers has steadily increased. It was mainly administrative and office buildings which were built with "Curtain Walls" (1894) and glass facades. It was followed by New York, Michigan and then Los Angeles, which started to build up. Before that time, it was rather churches and factory buildings in Europe that represented the tallest buildings in the world. The Chicago "Sears Tower" with a height of 442 m (1974) was after the completion, the tallest building in the world, replacing the "World Trade Centre" (417 m) in New York as the tallest skyscraper[2-4]. These heights were exceeded for a long time. However, not in the United States, where everyone would have expected it, but in the Arabian World and in Asia, arise now the tallest skyscrapers of our time. In Shanghai and Dubai, for example, you can find the "Shanghai Tower" and the "Burj Khalifa" with the heights of 632 m and 830 m, which are the second tallest and the tallest structure in the world[5]. But by the end of the decade, the "Burj Khalifa" will lose the title of the tallest building in the world, to the "Jeddah Tower" in Saudi Arabia. This is to be completed in 2021. Its planned height is around 1 000 m and has a spiral, needle-like shape.

These examples make clear how fast the development of skyscrapers is progressing and that today's dimensions, that were previously considered unimaginable, are achieved. With important former inventions like the passenger elevator and today's state of the art, there are hardly any limits to the realization of an extraordinary structure. The progressive development in the availability, processing and strength of building materials, as well as the high level of technology of the construction industry makes it possible to realize such construction projects.

However, not only on skyscrapers, but also on all high-rise buildings of modern times, more and more requirements are set. Especially in Europe, the development that is being used to build high-rise buildings is immense. Particularly in the areas of fire protection and energy efficiency, higher and higher standards are set, which the applicable laws and regulations must be built firstly.

2 Loads and Forces on Skyscrapers

There are three types of loads and forces generally:

2.1 Dead Load

Dead loads are the loads of the structure and fixed components. It is a permanent force that is relatively constant for an extended period of time. The force is also gravitational.

2.2 Live Load

Live loads are changing forces generated by mobile objects inside the building, such as people within the building or stock in a warehouse. Here the force is also gravitational.

2.3 Environmental Load

Environmental loads are forces acting on the building from its environment and may include wind, rain, earthquakes and temperature changes. The forces created can be either horizontal or vertical, positive or negative.

2.4 Vertical Forces

Dead loads and live loads contribute to the vertical forces on the structure of buildings. Compared to that, vertical loads are transferred from the floors to the columns and walls, and eventually to the soil or bedrock. At times, environmental loads also act vertically.

2.5 Horizontal Forces

Environmental loads contribute most of the horizontal forces acting on the structure of buildings, with loads from wind being the most common. Architects refer to these horizontal forces as shear forces. Adding cross bracing or shear walls can improve structural resistance to shear forces.

2.6 Internal Forces

The internal strength of the entire structure must be = or > the total forces applied on the building. The ability to withstand all forces depends on the structural components' dimensions and the solidity and elasticity of the material. Internal forces are compressive and tensile forces. According to Newton's Third Law, forces act in pairs. In structural terms, tensile force pulls a structural element apart while compressive force compresses it. If opposing forces are applied at different points, a structural element may become twisted. That is called torque[6].

3 Typical Structural Systems in Skyscrapers

3.1 Core and Outrigger Structure

The International Commerce Centre is built using a "Core and Outrigger" concept. The core at the centre of the building bears most of the vertical loads, while columns at the perimeter carry less weight and are thus smaller in dimension. Loads are transferred to the core through steel outriggers that balance the lateral forces on the whole building.

3.2 Steel

It is a common construction material for tall buildings and has good performance in withstanding compressive and tensile forces, as opposed to concrete's low tensile strength in compression. Steel bars can be used to reinforce concrete to add extra structural performance. But on the other hand it's relatively weak in fire-resistance.

3.3 Truss

Truss is a common structural element in architecture. Steel members are joined together into triangular shapes, which are able to resist external forces. After this, these triangles can form large truss systems that can span long distances[6].

4 Conclusions

The exploration of all these forces, loads and structural systems helps the humanity to build higher buildings in the future and will more and more reflect the character of a city. That is why it is more important that the design of the exterior facade and the building construction fit perfectly into the urban planning picture, as a skyscraper is clearly recognizable from far because of the enormous height.

Nevertheless, they are often an expression of power or a prestige object for the owner, builder or government. The symbolizing of economic strength is also a reason.

In addition, it should be a building, which is also suitable and energetic for the future. Another aspect is that many tourists are attracted to admire the biggest giants in the world.

References

[1] http://www.uni-kassel.de/upress/online/frei/978-3-89958-047-1.volltext.frei.pdf Access: 20.08.2018.

[2] http://www.theguardian.com/cities/2015/Apr/02/worlds-first-skyscraper-chicago-home-insurance-building-history, Access: 20.08.2018.

[3] http://www.elkage.de/src/public/showterms.php?id=8, Access: 20.08.2018.

[4] Johann N. Schmidt.: Wolkenkratzer: Ästhetik und Konstruktion, DuMont Buchverlag, Köln, Erstveröffentlichung 1991. Res., 42(1):11-14.

[5] http://www.diewolkenkratzer.de/wolken-kratzergeschichte.html, Access: 20.08.2018.

[6] Johann Eisele, Ellen Kloft (Hrsg.).: Hochhaus Atlas, Verlag Georg D. W. Callwey GmbH & Co. KG, Veröffentlichung 2002, Res., 87(1): 138-145.

(3) Surprises During Remediations

Laura Pelzer

Department of Civil Engineering, University of Magdeburg-Stendal

Abstract

In every profession, time management and coordination are important. It gets inconvenient when something unexpected and time-consuming comes up. In the building sector, as everywhere else, one must not despair but must act quickly and work on feasible solutions. Often something unplanned happens during the remediation of old buildings; for example water damage. These are often no trifle for buildings and also a financial burden. Early detection and scheduling of these problems, before the start of construction works and during the planning, is the basis for the success of the building project. Furthermore you will be spared unpleasant surprises.

Key Words

Remediations, Management and Coordinations, Profession

1 Introduction

Everywhere is currently under construction. Construction zones delay traffic if you want to go to university in the morning. Cranes are increasingly appearing in the skyline of cities. It's obvious that the construction industry is booming. However, not only modern glass boxes are built but also more and more remediations are being carried out.

Turning old into new is the motto.

In most cases, historical facade is largely preserved during renovation and the inner core is thoroughly touched. Nevertheless it's important to achieve today's high standards in order to be able to compete with the new buildings.

The fundamental renovation of a building is often much more demanding and complex than a new building. An old building hides many surprises that were not expected or planned. But why do the building owners make the effort to renovate anyway?

The charm of an old town villa from the baroque period or a water mill on a stream is often not only something for the eye, but is also very difficult to find soon after. Anyone can build new. But buildings, built in the old way and in the old style, are hardly made.

That's why I'm always happy when old buildings are preserved and the town and town's historical past are preserved.

So I was also able to accompany a

construction project in the Magdeburger Wissenschaftshafen, where the investor and builder decided to convert a large industrial storage facility into a modern office and laboratory building for the University of Magdeburg[1,2].

2 Surprises at the Old Building

As already mentioned, the old building often comes with unexpected surprises. Especially when demolishing, you will find all kinds of things under the groundwork and in the foundation. These were usually not considered in the planning, since they are not known and not noted in old plans. But exactly these additional works are reflected expensively in the supplements of the construction companies.

In order to avoid such additional costs, examinations are often ordered in the apron. Drilling into the foundation aims to determine the materials in the individual layers. The foundation walls are sampled to determine possible contaminations or other impurities, for example.

These early examinations were also carried out on my object. The building had been empty for years and hardly any repairs or refurbishments were carried out during this time. One could already see with the naked eye that the roof had some leaks and it rained in at several places.

Particularly affected were the exterior masonry and the associated masonry piers.

Roof drainage was carried out along the masonry piers. As these were defect, especially the piers were soaked. A contracted company should investigate and assess the damage in more details. They drilled holes from the masonry in the affected areas and examined them in the laboratory. It turns out that the piers at the worst points were saturated with up to 250 liters per m^3.

Since the building requires a dry and damage-free wall surface, a solution for soaking had been found quickly.

The commissioned company advised us to use infrared dark emitters because of the large amount of wall surface. These should be able to dry the masonry to the depth of 30 cm. If drying had been achieved in this area, the panels would have to be rearranged. The moisture in the cross section would then have time to come after and one would start again a drying interval. Until finally the masonry wall would be completely dry.

No time was planned for this in the existing construction schedule. Parallel drying during construction would also be a logistical nightmare.

When asked how many heating periods were needed and how long the complete drying process took, hardly any company could give us an answer. The internal masonry area is around 2,000 m^2. About 30% of this would have been dried. So it is not a small measure that could not be underestimated.

A second question for which a solution was needed was the renovation of the inside of the existing masonry.

There were several variants to choose from. For example, the use of restoration plaster or the application of a waterproofing slurry. The two most expensive variants included the stone repair of the entire surface and possibly the attachment of an acrylic disc.

3 Conclusions

I was lucky enough to be in this project and to learn, as I worked in the planning office alongside my studies. It has shown me that no matter how well one thinks to have planned, it can always happen that by such a measure, the whole construction project is endangered[3].

References

[1] Assmann Beraten und Planen, 2017. design department, implementation planning.

[2] picture of own documentation, 2018. a result of an object inspection, building condition documentation.

[3] BKG-Bannasch, 2015. website of the company recording image of IR-plates and their use.

(4)Supporting an Excavation Pit with Steel Piling

Laura Niens

Department of Civil Engineering, Magdeburg-Stendal University of Applied Sciences

Abstract

Building engineering is the application of engineering principles and technology to building design and construction. To erect and found physical structures including buildings, traffic facilities and line constructions, there is often a building pit necessary which is generally the space digged under the ground level.

The pit's closure needs to be built following a lot of rules. You can close a pit with embankments which you need a lot of space for. But if there is a lack of space for example in an inner-city area you can save this space by erecting a temporary building pit support system. There are a lot of different support systems. It differs between flexible systems like the so-called Berlin support system—which consists of steel beams and planks—or steel piling and inflexible systems such as bored pile wall or diaphragm wall. Out of that the steel piling is a very popular kind of supporting system which is basically a wall made out of steel profiles. Not only they are easy to build but they can also be more than just a building pit support system. In this paper you will get to know the use of them better.

Key Words

Pit, Piling, Berlin Support System

1 Introduction

A steel pile wall is used for supporting the soil mass laterally so that the soil can be retained at different levels on the two sides. Besides using the steel piling as a supporting system, it can also be used to function as a sealing against water or as an immobilisation of harmful substances by surrounding contaminated ground soil.

Steel pile walls are made out of single profiles which are normally out of steel but can also be vinyl, ferroconcrete or wood planks in exceptional cases. These single profiles are called sheet piles (Figure 1).

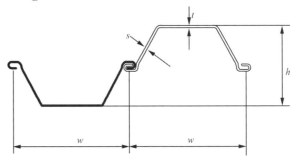

Figure 1　Profile of a sheet pile with height (h), width (w), thickness (t/s) - U-Profile as an example

There are some kinds of different profiles. The most common ones are U-profiles, Z-profiles and Omega-profiles. Their difference lies in whether the lock is situated on the web or flange.

The piles can be vibrated, rammed or pressed into the soil. Therefore, special construction machines with frame leaders are needed. With those kinds of machines, it is also possible to pull the sheet piles out of the soil again, so they can be reused(Figure 2).

Figure 2　Pile driver ram (Liebherr LRH 100) with a frame leader [1 - www.liebherr.com]

The single boards are interconnected with interlocking locks (tongue and groove system). That is how a continuous wall is formed. While installation, each board is guided in the lock of the last one rammed for form-fitted connection. To create a complete waterproof wall, the locks are either welded afterwards or inserted with flexible membrane lining.

The main domain is the security of building pits where there is no space for embankments or where there is a need of sealing against pressing water. Those are temporary security measures. It is also possible to construct in pits enclosed by steel boards underneath groundwater level by having a natural water-proof ground, creating an underwater concrete base or lowering the groundwater level with dewatering.

Besides temporary constructions, steel pilings can also be permanent when used as elements in hydraulic engineering or flood prevention.

2　Dimensioning

For a quick estimate the material is usually driven 1/3 above ground, 2/3 below ground, but this may be altered depending on the environment. The simplest case is the steel piling held in the ground by an appropriate large embedment depth which corresponds the static system of a cantilever.

Taller sheet pile walls will need a support structure—which need for example an inside horizontal waler as bearing—such as a tie-back anchor, or "deadman" placed in the soil a distance behind the face of the wall, that is tied to the wall, usually by a cable or a rod. Anchors are then placed behind the potential failure plane in the soil[1] (Figure 3, Figure 4).

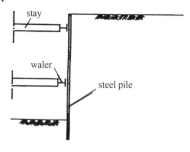

Figure 3　Two-layered stiffened system with waler and stay

Figure 4　Stiffened steel piling [2 - www.hk-ingbau.de]

The aim of the design of steel piles is the required depth of penetration, the maximum bending moment, the evidence of vertical forces and stress analysis (Figure 5).

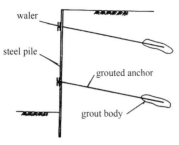

Figure 5　Double-anchored system with waler and grouted anchor with grout body

The design needs the requirement of the soil properties, the hydraulic pressure and the static system. That is why there is a foundation ground reconnaissance necessary which tests the drivability and locates possible obstacles. This includes drill-holes, cone penetration test and driving tests.

There are a lot of different methods of designing the steel pile walls. Classic methods for example are based on measuring the earth pressure and earth resistance and their distribution.

The dimensioning of steel pilings is based on purely static aspect. But the final choice of the profile, length, thickness and grade of steel depends on a lot of different influences such as difficulties in driving, constructional life span and possible corrosion and reusability respectively.

3　Conclusions

Even though the steel pile wall is comparatively a more expensive way of closing a building pit, it has a little cross section with a high stability which makes their drivability better.

They are extricable which makes them usable more often.

If the ground reconnaissance is well made and the dimensioning is truly based on the results, the failure probability is near zero because of the profile's high stability which is why steel pilings are used so often.

All in all, they are fully approved elements in different sectors in building industry.

References

[1] L. M. Zhang. Model and Application of Neural Network[M]. Shanghai: Fudan University Press, 1993.

(5) How to Make a High-speed Railway Tunnel Safe in Case of Emergency

Manuel Kempf, B. Sc.

Magdeburg-Stendal University of Applied Sciences Department of Water,
Environment, Construction and Safety Otto von Guericke University of Magdeburg
Institute of Instrumental & Environmental Technology

Abstract

The following article presents the recently opened German high-speed railway line VDE 8.1 in general and its safety conception in detail. The top speed of 300 kilometres per hour and the remarkable number of 22 tunnels and 29 bridges along the route make it necessary to reduce the risk of incidents as well as support rescue operations in case of an emergency.

To cope with the special circumstances in tunnel buildings, numerous safety measures were taken. Some examples are shown below. They all follow the aims to prevent an incident from occurring, to minimize the consequences of an incident and to support the rescue of passengers.

At first, the tunnel buildings themselves fulfil basic safety purposes by the way they were planned and constructed. Next, the technical building equipment of the tunnels contributes an important part to the rescue of passengers and emergency operations. The last presented point is the prepared concept for operations of emergency services and fire departments in case of an incident in one of the tunnels.

Key Words

TMD, Swing Vibration Control

1 Introduction

In 2017 the high-speed railway linking between Berlin and Munich started its regular operation. The so called "German Unity Transport Project 8" (VDE 8) is a part of the Trans-European Scandinavian-Mediterranean Transport corridor. It shortens the travel time between the two cities from formerly 6 hours to now 4 hours with a maximum speed of 300 kilometres per hour on 230 kilometres of newly built tracks. Another 260 kilometres were upgraded to allow a top speed of 200 kilometres per hour. Up to 130 freight and passenger trains per day and direction are planned to use the line, which is, in parts, automatically operated and controlled by the European Train Control System without any light signals[1].

The last part of the project to start its regular operation (VDE 8.1) runs from Erfurt in Thuringia to Ebensfeld in Bavaria. Especially this part of the line has to overcome a difficult topography with mountain ranges and many valleys across the Thuringian Forest. This leads to the large number of 22 tunnels (total length 41 kilometres) and 29 bridges (total length 12 kilometres). The whole line measures 107 kilometres, so nearly half of its length consists of engineering structures. The longest tunnel, the Blessberg Tunnel, is 8 314 metres long, and the longest bridge, the Ilmtal Bridge, measures 1 681 metres.

2 Tunnel Safety Considerations

The German Fire Protection Association developed an emergency scenario in 2001, originally for the high-speed line between Cologne and Frankfurt[2]. The assumed "incident in a tunnel" at VDE 8 starts with a passenger train that unexpectedly has to stop inside a tunnel. Reasons could be technical failures on the track or the train or a fire on the train. The involved train is around 400 metres long and on average carries 300 passengers. Up to 60 of them are injured and about 20 are not able to walk on their own, so they need help and have to be rescued.

The tunnel environment brings some special challenges to cope with in the emergency operations. First of all, the train has to be evacuated and several hundred passengers have to get through the tunnel to the outside. Fire, smoke and heat can expand much faster in the confined environment of a tunnel tube than outside or in a normal house. The length of the buildings and their location are making the emergency operation harder. In Thuringia, the railway line mostly runs through rural, remote and thinly populated areas of the Thuringian Forest. This increases the time required for the mainly voluntary emergency services to reach the tunnel buildings.

To make the best reaction to such emergency scenarios possible, safety measures are necessary to[3]:

(1) prevent an incident from occurring,

(2) minimize consequences of an incident,

(3) improve and support the rescue of passengers on their own and by emergency services and the fire department.

From the civil engineer's and safety engineer's point of view, possible categories for the adopted measures could be measures concerning:

• the tunnel building itself,

• the technical building equipment,

• the operation of fire department and emergency services.

Some chosen examples are presented in the following three sections.

3 The Building Construction

In the part running northward from Erfurt to Leipzig and Halle (VDE 8.2), the three tunnels are built as "two tube tunnels". The single tubes with one track in each are connected by cross-passages every 500 metres. In case of an incident, the parallel tube can be seen as a "safe area" in contrast to the involved one. Passengers will use it to leave the tunnel building. Emergency services can drive inside to prepare their operations as close as possible to the incident site.

In contrast to the double tubes at VDE 8.2, in the southern part from Erfurt to Ebensfeld (VDE 8.1), the line leads through tunnels with one double-track tube. Therefore a safe parallel tube in case of an emergency does not exist.

Passengers can leave the double-track tube through smoke- and fire-safe doors every 1 000 metres. The following emergency exits lead directly or via staircases to the outside. In longer tunnels, up to four exits are connected with a separate rescue tunnel (Figure 1).

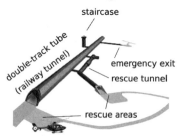

Figure 1 Principle sketch showing the possible emergency exits of tunnels at VDE 8.1(figure adapted from [F1]).

61 rescue areas with 1 500 square metres each are prepared outside the exits and around the tunnel portals. They serve as set-up space for incoming emergency services. Emergency operations to the inside of the tunnel will start from there, and passengers leaving the tunnel will get first aid care if necessary.

To protect the tunnel building, the passengers and the operating emergency units from the consequences of the fire heat, all building materials used inside the tunnel have to withstand a fire with temperatures of 1 200 degrees Celsius over a period of one hour [4].

4 Technical Building Equipment

Inside and around every tunnel building, many technical systems serve safety purposes and support emergency actions.

The first one is the so called "emergency braking override system". If a technical system on the train detects an irregularity or a passenger uses the emergency brake, the emergency stop can be overruled by the train driver while the train passes through a tunnel. The braking will then start when the train has left the disadvantageous position. This allows the train to stop at a more accessible part of the track. In most cases, the short amount of additional time will not influence a fire or a technical defect as much as the overweighting advantages of stopping outside will later help defend it.

If nevertheless a train stops inside a tunnel, passengers can leave the wagons over makeshift steps belonging to every train. Platforms are not installed inside the tunnel and the tracks are built on a slightly elevated ground, so the steps are necessary as escape walkways to overcome the vertical distance of approximately one metre between wagon and tunnel ground.

Once the passengers have left the train, emergency lights and escape signage help them to find the direction to the nearest emergency exits. Close to the tunnel walls, an elevated pavement makes it easier for the passengers to walk their way to the emergency exits. Additionally, handrails on both sides of the tube help to stay close to the walls even in darkness.

Before any emergency services can enter the tunnel and start operating, it has to be ensured that the overhead line system is separated from its electricity supply and then earthed. This safety measure is offered and supported by a remotely controlled system used for checking the current and the earthing of the line. As a part of the overhead line system inside the tunnel, it also belongs to the building's technical infrastructure.

The system's status is signalised by lamps on displays at every tunnel entrance.

Nearby a communication interface provides direct contact to a train dispatcher of the DB Netz AG. This subsidiary company of the Deutsche Bahn AG (German Railways) is the infrastructure manager of VDE 8 and therefore responsible for the support of emergency operations.

The supply of extinguishing water for the emergency services is secured by permanently filled cisterns at every exit. Pipes with regular hydrants to connect hoses are installed inside the tunnel. Electrical lines and power outlets provide power for the usage of electrical rescue gadgets such as additional lighting or hydraulic cutters.

Another part of the technical building equipment are antennas and amplifiers for the emergency radio communication system TETRA (Terrestrial Trunked Radio).

5 Operationg of Emergency Services

In the early operation stage, the most important step for emergency services is to locate the precise incident spot in the tunnel. This is done by exploration units starting at every exit. Furthermore they will find the first injured passengers and mark their position for later rescue. Now the operation differs from the standard procedures at a normal fire or technical support. Due to the special conditions inside the tunnel, the main principle is "extinguishing to rescue"[5]. Cooling the building's structure protects it from structural damage caused by the heat, which would endanger the rescue. The expansion of smoke and heat is decreased so the rescue of the estimated 20 injured passengers not able to walk is made possible.

The emergency concept for VDE 8.1 involves up to 700 volunteer fire fighters and 300 volunteers from other emergency services across the Federal States of Thuringia and Bavaria.

A unique feature is the detailed tactical and organisational preparation. For each tunnel every fire department unit has preplanned places to set up and prepare, tasks to do and measures to take. The only variable is the tunnel in which the incident happened.

This allows a standardised training

for every involved emergency unit and the operation of qualified tunnel operation units all along the high-speed railway line from Erfurt to Ebensfeld [6].

6　Conclusions

Hopefully an incident like the one assumed in the shown scenario is never going to happen. In the end risks can never be fully eliminated. The best way of reaction is preparation.

The construction of the building, its technical equipment and the organisation of operation in case of an emergency are prepared in a way that the people involved in the planning and those responsible for the safety consider as the best support for the rescue.

Last but not least, steady exercising of emergency procedures and the concept is needed for training of the emergency units. This helps find the best practices and improve the whole presented safety and emergency system for the high-speed railway tunnels on the route VDE 8.1.

References

[1] DB Netz AG, Regional Unit South East, VDE 8 Project, 2016. The largest rail construction site in Germany. Leipzig, Germany.

[2] Vereinigung zur FÖrderung des Deutschen Brandschutzes e. V. (German Fire Protection Association), 2000. Empfehlungen zur Schadensbekämpfung bei Brand und Kollision von Reisezügen in Tunnelanlagen der Deutschen Bahn AG durch Öffentliche Feuerwehren. In: Mitteilungen der VFDB 3/2001, pp. 145-149.

[3] Kruse, 2016. Sicherheit in Eisenbahntunneln. In: BrandSchutz 8/2016, Stuttgart, Germany, pp. 577-583.

[4] Eisenbahn-Bundesamt (German Federal Railway Authority), 2008. Richtlinie Anforderungen des Brand- und Katastrophenschutzes an den Bau und den Betrieb von Eisenbahntunneln.

[5] Wagner, Stielow, 2016. Gefahrenabwehr in Bahntunneln auf der Neubaustrecke VDE 8.1. In: BrandSchutz 8/2016, Stuttgart, Germany, pp. 584-593.

[6] Kehsler, 2016. Thüringen: Ausbildung für die Tunnelbasiseinheiten des》VDE 8.1《. In: BrandSchutz 8/2016, Stuttgart, Germany, pp. 594-595.

(6) Research on TMD Parameters for Swing Vibration Control of Suspended Structures

Wang Hao, Zhang Chunwei

Department of Civil Engineering, Qingdao University of Technology

Abstract

The effect of Tuned Mass Damper (TMD) parameters on the control of suspended structures under in-plane swing motion mode is studied in this paper. Analysis shows that the principles of TMD control and vibration based monitoring by accelerometer for in-plane swing motion mode are unified. Results show that only the normal TMD is valid in control of in-plane swing motion mode of suspended structures.

Key Words

TMD, Swing Vibration Control

1 Introduction

TMD (Tuned Mass Damper) is a passive dynamic vibration absorber which has been applied widely in the domain of structural vibration control. Traditional TMD is made up of mass, spring and damping[1]. Most of the studies of TMD concentrate on control of linear motion of structures such as vertical vibration of bridges, lateral vibration of high-rise buildings. Torsional vibration exists widely in reality: vibration of long span bridges under the wind load contains torsional vibration component[2-4], eccentric buildings under the earthquake load will have torsional motion[5,6], and the in-plane swing motion mode of suspended structures can be regarded as a torsional motion around suspension points[7].

2 In-plane Swing Motion Mode Control With Normal TMD

2.1 Modeling Development

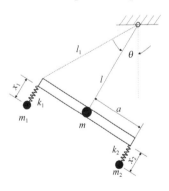

Figure 1 Simplified calculation model for the suspended structure with double normal mass-spring oscillators

For double TMDs, in order to further explore the control performance of mass-spring oscillator, another two simplified calculation models are developed. The first model is a pendulum with double normal spring-mass oscillators as shown in Figure 1. The system is considered to have three degrees of freedom: swinging angle (denoted by θ) and stroke displacements of mass-spring oscillator relative to the pendulum mass (denoted by x_1 and x_2). Characteristics of two oscillators are identical and several parameters are defined: $\mu_m = \dfrac{m_1}{m} = \dfrac{m_2}{m}$ (1) is the mass ratio, $w_2 = \sqrt{\dfrac{k_1}{m_1}} = \sqrt{\dfrac{k_2}{m_2}}$ (2) is the natural frequency of mass-spring oscillator, $\beta = \dfrac{\omega_2}{\omega_1}$ is the frequency ratio of mass-spring oscillator to pendulum.

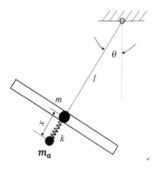

Figure 2 Simplified calculation model for the suspended structure with single normal mass-spring oscillator

The total kinetic energy and potential energy of this system (Figure 1) can be expressed as

$$T = \frac{1}{2}ml^2\dot\theta^2 + \frac{1}{2}m_1\left[(\dot x_1 + a\dot\theta)^2 + (x_1+l)^2\dot\theta^2\right]$$
$$+ \frac{1}{2}m_2\left[(\dot x_2 - a\dot\theta)^2 + (x_2+l)^2\dot\theta^2\right] \quad (3)$$

$$U = \frac{1}{2}k_1 x_1^2 + \frac{1}{2}k_2 x_2^2 + mgl(1-\cos\theta) +$$
$$m_1 g \left\{ \begin{matrix} -x_1\cos\theta - \\ [l_1\cos(\varphi-\theta) - l] \end{matrix} \right\} + m_2 g$$
$$\left\{ -x_2\cos\theta + [l - l_1\cos(\varphi+\theta)] \right\} \quad (4)$$

Where φ is the angle between l and l_1.

Based on the Lagrangian principle, the equation of the motion of pendulum with double normal mass-spring oscillators system can be achieved as

$$m_1(a\ddot\theta + \ddot x_1) + k_1 x_1 - m_1(x_1+l)\dot\theta^2$$
$$= m_1 g\cos\theta \quad (5)$$

$$m_2(-a\ddot\theta + \ddot x_2) + k_2 x_2 - m_2(x_2+l)$$
$$\dot\theta^2 = m_2 g\cos\theta \quad (6)$$

$mgl\sin\theta + m_1 g[-l_1\sin(\varphi-\theta) + \sin\theta x_1] + m_2 g[l_1\sin(\varphi+\theta) + \sin\theta x_2] + ml^2\ddot\theta + m_1[(2l\dot x_1 + 2x_1\dot x_1) + a\ddot x_1 + (l_1^2 + 2lx_1 + x_1^2)\ddot\theta] + m_2[(2l\dot x_2 + 2x_2\dot x_2) - a\ddot x_2 + (l_1^2 + 2lx_2 + x_2^2)\ddot\theta] = 0 \quad (7)$

$\dot\theta, \dot x_1, \dot x_2, \ddot\theta, \ddot x_1, \ddot x_2$ are velocity and acceleration along each degree of freedom.

For single TMD, the second model is a pendulum with single normal mass-spring oscillator as shown in Figure 2. The system is considered to have two degrees of freedom: swinging angle (denoted by θ) and stroke displacement of mass-spring oscillator relative to the pendulum mass (denoted by x).

The total kinetic energy and potential energy of this system (Figure 2) can be expressed as

$$T=\frac{1}{2}ml^2\dot{\theta}^2+\frac{1}{2}m_a\left[\dot{x}^2+(x+l)^2\dot{\theta}^2\right] \quad (8)$$

$$U=\frac{1}{2}kx^2+mgl(1-\cos\theta)+m_ag\left[-x\cos\theta-(l\cos\theta-l)\right]$$

Based on the Lagrangian principle, the equation of the motion of pendulum with single normal mass-spring oscillator system can be achieved as

$$m_a\ddot{x}+kx=m_1(x+l)\dot{\theta}^2+m_a g\cos\theta$$

$$mgl\sin\theta+m_a g(l+x)\sin\theta+ml^2\ddot{\theta}+m_a[2(l+x)\dot{x}\dot{\theta}+(l+x)^2\ddot{\theta}]=0$$

$\dot{\theta},\dot{x},\ddot{\theta},\ddot{x}$ are velocity and acceleration along each degree of freedom.

2.2 Numerical Analysis

The $\mu_m=2.5\%$, $\beta=1$ and 2, and other constants and parameters are the same as section 1.1. Solving equations (5) (6) (7) based on Simulink and the results are shown in Figure 3. The results show that no matter $\beta=1$ or 2 structural angle amplitude decreases and increases alternatively from time to time, which means the energy is transferred between the suspended structure and the mass-spring oscillator in time domain analysis and the double normal mass-spring oscillators are effective in control of in-plane swing motion mode of the suspended structure. But there is a different control principle between $\beta=1$ and 2. When $\beta=1$, directions of the motion of two mass-spring oscillators are reversed as shown in Figure 3(a). As a result, two reversed forces produced by double mass-spring oscillators can control the swinging of the pendulum. When $\beta=2$, the direction of the motion of double mass-spring oscillators is the same as shown in Figure 3(b) and the system behaves like a spring pendulum when $\beta=2$.

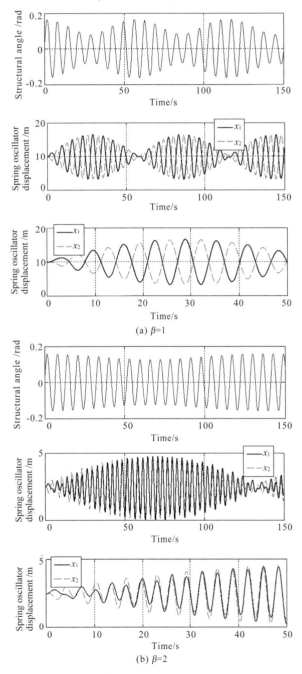

Figure 3 Time history of the suspended structure with double normal mass-spring oscillators in different frequency ratios

The $\mu_m = 5\%$ and other constants. Solving equations (10) (11) based on Simulink and the results are shown in Figure 4. The results show that only when $\beta=2$ the single normal mass-spring oscillator is effective in control of in-plane swing motion mode of the suspended structure as shown in Figure 4 (b) and when $\beta=1$ the mass-spring oscillator is invalid as shown in Figure 4 (a).

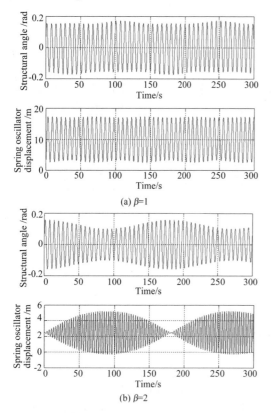

Figure 4 Time history of the suspended structure with single normal mass-spring oscillator in different frequency ratios

3 Conclusions

In order to explore the control and monitoring principles of in-plane swing motion mode, basic equations of the motion of simplified calculation model of in-plane swing motion mode of suspended structures are analyzed and some conclusions are achieved as following:

(1) Equations of motion of the suspended structure with normal TMD are established to explore the control performance of normal TMD as the tangential TMD is invalid in control of in-plane swing motion mode based on references. Numerical studies show that normal TMD is effective in control of in-plane swing motion mode no matter in double or single TMD condition. The control performance of tangential and normal TMD is determined by characteristics of acceleration along the tangential and normal directions of in-plane swing motion mode.

(2) Numerical results show that tuning principles in double and single normal TMDs condition are different. In double normal TMD condition, TMD is both effective when the frequency ratio $\beta=1$ and 2. But the work mechanism in two frequency ratio conditions can be classified into two modes: two reversed control forces control mode when $\beta=1$ and internal resonance of spring pendulum control mode when the $\beta = 2$. Single normal TMD can be effective only when $\beta=2$ and the work mechanism is the same as internal resonance of spring pendulum control mode.

References

[1] Ormondroyd J, Den Hartog JP. The theory of the dynamic vibration absorber. Transactions. ASME APM-50-7,1928; 9-22.

[2] Nobuto J, Fujino Y, Ito M. Study on the effectiveness of TMD to suppress a coupled flutter of bridge deck[C]// Doboku Gakkai Rombun-Hokokushu/ Proceedings of the Japan Society of Civil Engineers. 1988 (398 pt I-10): 413-416.

[3] Gu M, Chang C C, Wu W, et al. Increase of critical flutter wind speed of long-span bridges using tuned mass dampers[J]. Journal of Wind Engineering and Industrial Aerodynamics, 1998, 73(2): 111-123.

[4] Chen Airong, Xiang Haifan. Vortex-Induced Torsional Vibration Control of Cable-Stayed Bridges[J]. Journal of Tongji University (natural science), 1994, 22(4): 487-492. (in Chinese)

[5] Li Chunxiang, Han Chuanfeng. Optimum placements of multiple tuned mass dampers (MTMD) for suppressing torsional vibration of asymmetric structures[J]. Earthquake Engineering and Engineering Vibration, 2003, 06: 149-155.

[6] Fan Changlong. MTMD for torsional vibration control of tall building structures under seismic actions[D]. Tongji University, 2007.

[7] Zhang Chunwei, Xu Huaibing, Li Luyu, Ou Jinping. Structural pendulum vibration control methods based on tuned-rotary-inertia-damper (I): parametric impact analysis and bench-scale model tests[J]. Control Theory & Applications, 2010, 09: 1159-1165.

(7) Research on the Vibration Response of Subway with Disk-spring and Thick Rubber Isolator

Zhao Hanzhu, Sui Jieying, Li Yao

Qingdao University of Technology

Abstract

In view of the phenomenon of vibration caused by subway trains, a new type of disk-spring and thick rubber isolator has been proposed. A residential building along the 2nd line of Suzhou Metro from Xinjiaqiao Station to Panli Road Station was taken as the research object. The system of isolation was established with ANSYS and the wave of vibration simulated by subway was applied on the superstructure. The process was analyzed in this model system. And the results show that this isolator can obviously reduce the vibration response caused by subway and will be widely applied in engineering.

Key Words

disk-spring and thick rubber isolator, vibration of subway, time-history analysis, acceleration and displacement of vibration

1 Concept Design of the Isolator

1.1 Concept of the isolator

The device of the disk-spring and thick rubber isolator was shown in Figure 1. It was composed of three steel boards, disk-spring (three layers), thick rubber and bolts. It was divided into two parts—disk-spring and thick rubber. The direction of vertical load was board, disk-spring, thick rubber, and at last, board. When subjected to the vertical vibration, the disk-spring and thick rubber were compressed and deformed to absorb energy.

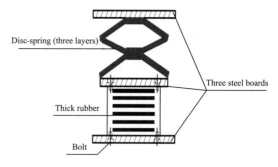

Figure 1　Disc-spring and thick rubber isolator

1.2 Design of the Isolator

The elastic modulus of the disk-spring was 2.06×10^5 MPa and the Poisson's ratio was 0.3. The maximum force of the vertical column was 525kN. The load was considered between 500kN and 550kN without bearing surface. The size of the isolator was $D \times d \times t(t') \times H_0 =$

280 mm×142 mm×16 mm×22 mm, which had the form of combination and four disks for each stacking group. The formulas of single disk-spring force were as following:

$$F = \frac{F_z}{n} = \frac{525}{4} = 131.25 \text{ kN} \quad (1)$$

$$C = \frac{D}{d} = \frac{280}{142} = 1.97 \quad (2)$$

$$K_1 = \frac{1}{\pi} \cdot \frac{(\frac{C-1}{c})^2}{\frac{C+1}{C-1} - \frac{2}{\ln C}} = 0.69 \quad (3)$$

$$F_c = \frac{4E}{1-\mu^2} \cdot \frac{h_0 t^3}{K_1 D^2} \cdot K_4^2 = \frac{4 \times 2.06 \times 10^5}{1-0.3^2} \times \frac{6 \times 16^3}{0.69 \times 280^2} \times 1 = 411 \text{ kN}$$

The stiffness of the disk-spring had been related to $\frac{h_0}{t} = 0.375$ and $\frac{h_0}{t} = 0.32$. According to the value of $\frac{h_0}{t}$, $f = 0.32 h_0 = 0.32 \times 6 = 1.92$ mm, the stiffness of the disk-spring was[1]:

$$F' = \frac{4E}{1-\mu^2} \cdot \frac{t^3}{K_1 D^2} \cdot K_4^2 \cdot \{K_4^2 \cdot \{K_4^2 \left[(\frac{h_0}{t})^2 - 3\frac{h_0}{t}\frac{f}{t} + 1.5(\frac{f}{t})^2\right] + 1\}$$

$$= 70\,428.14 \text{ N/mm} \quad (4)$$

When the friction was ignored, the number of the combined disk-springs was:

$$f_z = i \cdot f \quad (5)$$

$$i = \frac{f_z}{f} = \frac{6}{1.92} = 3.125 \quad (6)$$

The thick rubber had vibration reduction capability[2] to some extent on basis of traditional thin rubber via increasing the thickness of single one. The size of the selected thick rubber was $D \times A \times T_r \times N_r = 280$ mm×61 575 mm×4 mm×5 mm, and the stiffness was:

$$E_{cb} = \frac{E_c E_b}{E_c + E_b} = \frac{50 \times 2\,000}{2\,050} = 49 \text{ MPa}$$

$$K_R V = \frac{E_{cb} A}{h} = \frac{49 \times 61\,575}{22} = 136.5 \text{ kN/mm}$$

The stiffness of the combined isolators was:

$$K_v = \frac{1}{\frac{1}{F'} + \frac{1}{K_R V}} = \frac{1}{\frac{1}{70.428} + \frac{1}{136.5}}$$

$$= 46.458 \text{ kN/mm} \quad (7)$$

In summary, we selected four three-layer combination disk-springs without surface, the size of each one was $D \times d \times t(t') H_0 = 280$ mm×142 mm×16 mm×22 mm. The size of the thick rubber was $D \times A \times T_r \times N_r = 280$ mm×61575 mm×4 mm×5 mm. We combined them. The designs of the steel boards and bolts had been referred to the calculation[3] as well.

2 Model establishment

The superstructure along the metro was a reinforced concrete building with the height of 2.8 m and had 6 floors. The size of the cross-section of the column was 400 mm×400 mm and was made of concrete C40. The reinforced steel of the column was HRB400, and the stirrup was HPB335. The concrete of the slab and roof was C35, and the steel was HRB400. The details were shown in Figure 2. The distance between the subway and the adjacent building was 8.2 m, as shown in Figure 4. The material parameters were shown in Table 1.

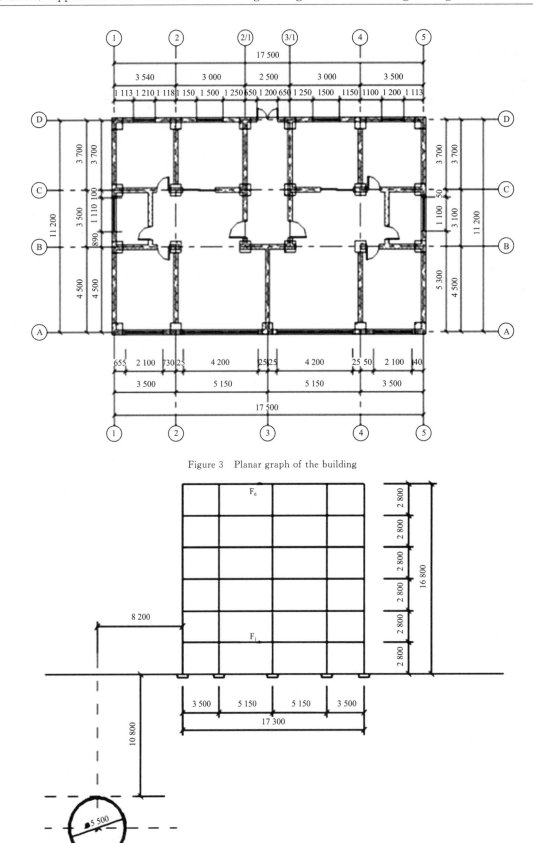

Figure 3 Planar graph of the building

Figure 4 Distance between the subway and the building

(7) Research on the Vibration Response of Subway with Disk-spring and Thick Rubber Isolator

Table 1   Material parameters

Name of material	Elastic modulus E(MPa)	Density ρ(kg/m^3)	Poisson's ration υ
Concrete C35	3.15×10^4	2 400	0.2
Concrete C40	3.25×10^4	2 400	0.2
Stirrup HPB335	2.1×10^5	7 850	0.3
Reinforced Steel HRB400	2.0×10^5	7 850	0.3

In this paper, we had selected SOLID65 as reinforced concrete element, PLANE42 as floor, COMBIN14 as vertical stiffness of the disk-spring isolator, and COMBIN40 as x and y horizontal directions[4]. The form of the structure was integral, as shown in Figure 5.

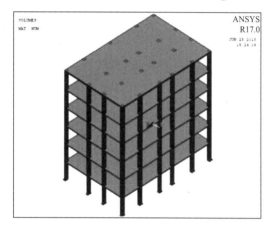

Figure 5   Model of the superstructure

3 Simulation Results

The vertical vibration was more significant than horizontal one when studying subway vibration[5]. The simulated acceleration curve was as following, as shown in Figure 6.

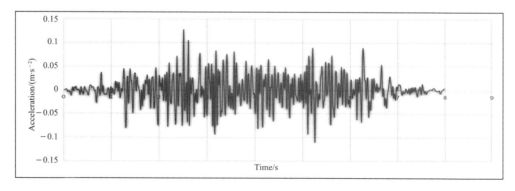

Figure 6   Simulated acceleration of the subway

3.1 Vibration Results Without Isolator

The force was applied to the structure without isolator. Point F_1 ($x=7\ 600$, $y=7\ 300$, $z=2\ 800$) of the ground floor and F_6 ($x=7\ 600$, $y=7\ 300$, $z=16\ 800$) of the top floor of the superstructure were selected as the representative points.

Then we could obtain the corresponding acceleration and displacement curve before isolation.

It could be seen from Figure 7 to Figure 10:

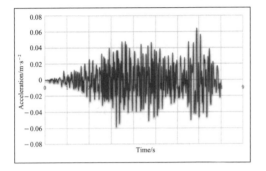

Figure 7　Acceleration curve of F_1

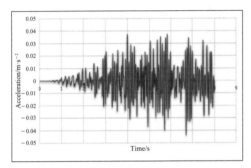

Figure 8　Acceleration curve of F_6

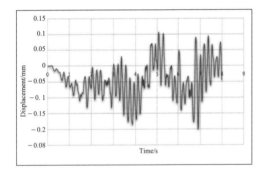

Figure 9　Displacement curve of F_1

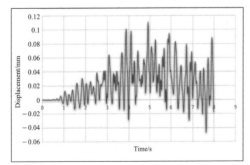

Figure 10　Displacement curve of F_6

The simulated acceleration and displacement curves of F_1 and F_6 were not exactly consistent, and the attenuation process of the F_6 acceleration curve was also shorter than F_1. As for displacement, the top and bottom ones changed in opposite direction. From the numerical point of view, the vibration induced by the subway was impaired by the frame. According to the Vibration Standards of Urban Regional Environmental[6] of China, the maximum of acceleration during the day should not exceed 31.6 mm/s², and not exceed 22.3 mm/s² at night. At the same time, if the superstructures were within 10 m far from the subway, they would be greatly affected[7] by the vibration. Therefore, we should adopt necessary method to reduce vibration and noise.

3.2　Vibration Results with Isolator

The simulated excitation of acceleration was applied to the superstructure with isolation. The acceleration and displacement curves of F_1 and F_6 after vibration isolation were shown in the following figures.

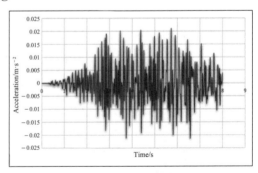

Figure 11　Acceleration curve of F_1

4　Conclusion

(1) The process of simulation was

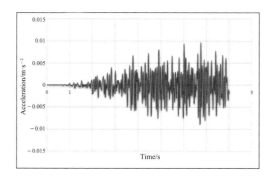

Figure 12　Acceleration curve of F_6

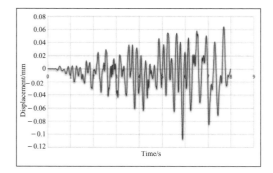

Figure 13　Displacement curve of F_1

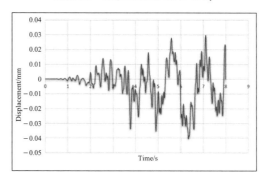

Figure 14　Displacement curve of F_6

conducted to verify the vibration reduction capability of the disk-spring and thick rubber isolator. It was obvious that the vibration reduction of the superstructure along the subway was effective—the peak of the acceleration of the bottom decreased 67.2%, the mean value decreased 64.3%. The maximum of the displacement decreased 46.3%, and the mean value decreased 65.5% with vibration isolation. Accelerations of the top floor had little change in the first 2 s and there was no obvious attenuation process with vibration isolator—the peak of the acceleration reduced 77.3%, the mean value reduced 75.0%, the maximum of the displacement reduced 64.0%, and the mean value reduced 64.0%.

(2) The vibration waves induced by the subway without isolator were attenuated by the frame, but still did not meet the standards. While the maximum of the acceleration was 21 mm/s² after adopting isolator, which met the standards. This isolator could obviously reduce the vibration response caused by subway and would be widely applied in engineering.

References

[1] GB/T 1972-2005, disk-spring [S].

[2] Sheng Tao, Li Yaming, Zhang Hui, Shi Weixing, Yang Yue. Study on Vibration Isolation of Thick Rubber Isolators in Adjacent Buildings of Subway [J]. Journal of Building Structures, 2015, 36(02): 35-40.

[3] General Administration of Quality Supervision, Inspection and Quarantine of the People's Republic of China. China National Standardization Administration Committee. GB 20688.3-2006 Rubber Isolators: Part 3: Building Isolation Rubber Isolators [S]. Beijing: China Standard Press, 2006.

[4] Liu Pingping, Sui Jieying, Wang Xuepeng. Study on Vibration Isolation of Existing Buildings under Subway Environment[J]. Journal of Qingdao University of Technology,2014,35(03):7-10+39.

[5] Yan Weiming, Zhang Wei, Ren Wei, Feng Junhe, Nie Wei, Chen Jianqiu. The Measurement and Propagation Law of Vibration Induced by Subway Operation[J]. Journal of Beijing Polytechnic University, 2006（02）:149-154.

[6] GB 10070-88 "Urban Regional Environmental Vibration Standards"[S]. Beijing: China Building Industry Press, 1988.

[7] Zhang Jun, Rui Wenjian. The Propagation Law of Vibration of Low Track Structure[J]. Journal of Qingdao University of Technology,2016,37(05):19-25.

(8) Carbon Source Supplement Strategies Make Difference in a Semi-centralized Black Water Treatment System

Li Siyu[1], Yin Zhixuan[1], Bi Xuejun[1], Yang Benliang[1], Huang Kai[2]

1 Qingdao University of Technology, 11 Fushun Road, Qingdao 266033, P. R. China,
(E-mail: 13256876631@163.com, yzxqut@163.com, 13969850081@126.com, 18766228310@163.com)

2 The 7th Design Institute of Shanghai Municipal Engineering Design Institute (Group) Co., LTD, Qingdao 266033, P. R. China, (E-mail: huangkai7301@126.com)

Abstract

Black water contains high content of nitrogen required external carbon source for denitrification. In this study, carbon source supplement strategies for pre-denitrification or post-denitrification were compared in a pilot-scale semi-centralized black water treatment system. In the post-denitrification carbon source addition, the TN removal rate increased from 78.9% to 88.8%. Under the same operating conditions, the pre-denitrification carbon source addition could increase the denitrification rate, but the TN removal rate was only 79.6%, and denitrification effect was not obvious. The results showed that carbon source supplement for post-denitrification was much more effective on the efficiency of denitrification and dephosphorization.

Key Words

Black Water Contains, Carbon Source Supplement

1 Introduction

Black water is wastewater from toilet and kitchen. The chemical oxygen demand (COD)/TN ratio is usually low, thus resulting in limited carbon source for nitrogen removal (Zhang et al., 2017). In the black water system biochemical pool, the Total Nitrogen (TN) concentration of influent water was high and the carbon source was limited and the denitrification effect was unstable. With the decrease of the sludge concentration, the effect of simultaneous nitrification and denitrification was weakened, and system effluent Nitrate Nitrogen (NO_3-N) concentration was increased to 22.5~26.5 mg/L, so it was necessary to improve the system's denitrifiction ability by adding carbon source (Hocaoglu and Orhon, 2013). Black water treatment system carbon source dosing points are the pre-denitrification and post-denitrification addition.

This test by experimentally compared the nitrogen and phosphorous removal effect of the without carbon source addition and pre-denitrification carbon source addition and post-denitrification to judge which dosing carbon source method was better (Hocaoglu et al, 2013).

2 Material and Methods

In this study, the influent was black water after primary sedimentation treatment. The average total COD (TCOD) concentration was 580 ± 85 mg/L. The TN and NH_4^+-N concentrations were 116 ± 12 mg/L and 93 ± 25 mg/L respectively. And the average TP concentration was 12 ± 4 mg/L.

Carbon source supplement strategies were compared in a full-scale semi-centralized black water treatment system (Qingdao, China). As shown in Figure 1, the system consisting of a series of Anoxic/Anaerobic/Aerobic reactors was followed by membrane biological reactor (MBR). The working volume of MBR reactor was 96 m^3. The average influent quantity of the system was 335 m^3/d with mixed liquor recirculation ratio of 400% during the test. The dissolved oxygen (DO) concentrations in the aerobic zones and MBR were set at $1.0 \sim 1.5$ O^2 mg/L and $4.5 \sim 5.0$ O^2 mg/L respectively. During the test, the operational temperature of the system was $14.2 \sim 15.5$ ℃.

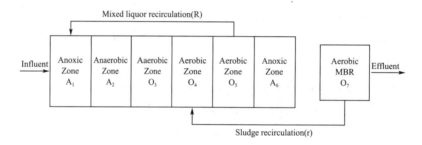

Figure 1 The flow diagram of the black water treatment system (A1: pre-denitrification zone; A6: post-denitrification zone).

As shown in Table 1, the system was operated without carbon source supplement (Stage Ⅰ and Ⅲ), with carbon source supplement into reactor A6 for post-denitrification (Stage Ⅱ) and with carbon supplement into reactor A1 for pre-denitrification (Stage Ⅳ). Sodium acetate was used as carbon source with concentration of 50 mg chemical oxygen demand (COD)/L mixd liquid.

Table 1　　　　The experiment data of carbon source addition

Stage	Ⅰ	Ⅱ	Ⅲ	Ⅳ
Time (Day)	0—5	6—31	32—41	42—47
Carbon source dosing place	-a	A6	-a	A1
Carbon source dosing concentration (mg COD/L)	0	50	0	50
TN removal efficiency/%	78.7 ± 3	87.2 ± 3	80.9 ± 3	79.6 ± 3
COD removal efficiency/%	97.3 ± 1	97.4 ± 1	97.5 ± 1	96.5 ± 1

a Without carbon source

The liquid samples were collected for chemical analyses. All conventional parameters such as COD, TN, NH_4^+-N, NO_3^--N and PO_4^{3-}-P were analyzed according to standard methods (APHA, 2005).

3 Conclusions

As shown in Figure 2(A), when the system was operated without carbon source supplement at stage Ⅰ (Day 1~Day 6), the average TN removal efficiency of the system was 78.7%, resulting in 24.7 mg TN/L in the system effluent. When the carbon source was added to reactor A6 for post-denitrification (stage Ⅱ, Day 7~Day 31), the TN removal efficiency increased significantly to 88.8%, maintaining an average effluent TN concentration of 13.5 mg/L. After stop adding carbon source (stage Ⅲ, Day 32~Day 41), the nitrogen removal rapidly recovered to 80.9%. From the Day 42 (stage Ⅳ), carbon source was added to reactor A1 for pre-denitrification but no promotion in nitrogen removal was observed. Furthermore, during the entire test, the COD removal was > 97.0% [Figure 2(B)], which was not influenced by carbon source addition.

Nitrogen and phosphorous transformations with/without carbon source supplement were further investigated in batch tests. As shown in Figure 2(A), it was observed that without carbon source addition in the black water treatment system, nitrogen was mainly removed in the first 150 min with TN removal rate of 6.4 mg N/(L · h), and influent nitrate was mostly reduced by denitrification. Once carbon source was supplemented after aeration stage (at the 350th min) for post-denitrification, TN was further removed by 13.2 mg/L [Figure (2)B]. That was, 3.8 mg COD was consumed with 1 mg TN removal by post-denitrification.

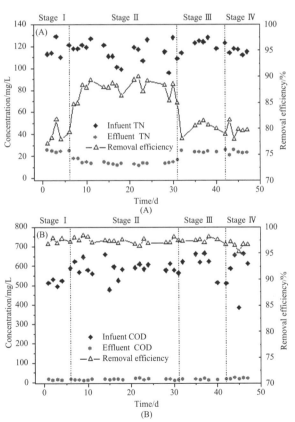

Figure 2 TN and COD removal from black water in the continuous-flow system with/without carbon source supplement for denitrification.

References

[1] Zhang, J., Chen, H., Tang, X. and Dai, X. (2017), Treatment of single and multi-AO-SBR process on black water. Chinese Journal of Environmental Engineering (in Chinese), 11(3), 1409-1416.

[2] Hocaoglu, S. M. and Orhon, D., Particle size distribution analysis of chemical oxygen demand fractions with different biodegradation characteristics in black water and gray water. Clean-Soil. Air. Water. 2013, 41(11), 1044-1051.

[3] Hocaoglu, S. M., Atasoy, E., Baban, A., Insel, G. and Orhon, D., Nitrogen removal performance of intermittently aerated membrane bioreactor treating black water. Environmental Technology, 34(19), 2013, 2717-2725.

(9) Influence of Reinforcing Ring Parameters on Seismic Performance of Rectangular Steel Tubular Column-H Shaped Steel Beam Joint

Zhang Zhipeng, Xu Peizhen, Li Qiming
School of Civil Engineering, Qingdao University of Technology

Abstract

The rectangular steel tubular column-H shaped steel beam joint with external reinforcing ring has excellent mechanical properties. The seismic performance by different parameters on this type of joint is investigated by Software ABAQUS, which includes the ratio of the length of the external reinforcing ring along the beam side to the width of the beam flange, the ratio of the thickness of the external reinforcing ring to the thickness of the beam flange. The results show that this type of joint has excellent seismic performance. What is more, the variation of these two parameters has significant influence on the failure mode, hysteretic behavior, skeleton curve, ultimate bearing capacity and stiffness degradation of the joints.

Key Words

Rectangular Steel Tubular Column, External Reinforcing Ring Parameter, FEM Analysis, Seismic Performance, Hysteretic Performance

1 Introduction

Previous studies show that the rectangular steel tubular column-H shaped steel beam joint with external reinforcing ring has excellent mechanical property[1-4]. Based on the previous research, the seismic performance of this type of joint is investigated by Software ABAQUS, which includes the ratio of the length of the external reinforcing ring along the beam side to the width of the beam flange, the ratio of the thickness of the external reinforcing ring to the thickness of the beam flange.

2 Finite Element Model of Joint

2.1 Finite Element Model and Parameters

Figure1 shows the constraint conditions of the finite element model(FEM). The contacting surface is defined by selecting the tie (binding constraint). The height of the model from the bottom fixed point to the vertical loading point is 3 000 mm, and the length of the unilateral beam from the outer edge of the column to the end of the beam is 2 000 mm[5]. The beam section and column section sizes are 600 mm×200 mm×14 mm ×10 mm and 450 mm×350 mm×20 mm respectively. According to the relevant code[6-9], the design of the reinforcing ring is carried out according to the 9-degree seismic fortification requirements. The reinforcing ring size of the reference model is shown in Figure 2. Based on the reference model, the non-dimensional parameters are designed by changing the length of the external reinforcing ring along the beam side ($B=150, 200, 250$ and 300 mm) and the thickness of the external reinforcing ring ($t_d = 12, 14, 16$ and 18 mm), which includes the ratio of the length of the external reinforcing ring along the beam side to the width of the beam flange ($B/b_f = 0.75, 1, 1.25$ and 1.5), the ratio of the thickness of the external reinforcing ring to the thickness of the beam flange ($t_d/t_f = 1.2, 1.4, 1.6$ and 1.8). The models were SJ-N1-0.75, SJ-N1-1, SJ-N1-1.25, SJ-N1-1.5 and SJ-N2-1.2, SJ-N2-1.4, SJ-N2-1.6, SJ-N2-1.8.

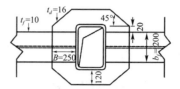

Figure 2 Details of external reinforcing ring

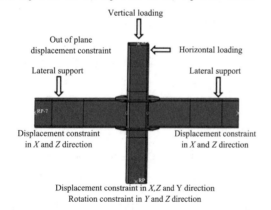

Figure 1 Finite element model and constraint condition

2.2 Material constitutive relations and loading system

The beams, columns and reinforcing rings are made of Q235B steel while the material properties of steel were shown in Table 1. The horizontal reciprocating load is applied at the end of the column, and the vertical load is applied which is equivalent to 0.2 axial compression ratio at the top of the column. The loading system was set according to the American AISC seismic code[10], taking the inter-story displacement angle R as the control parameter, and each cycle was loaded two times. The first six loading stages were: 0.375%, 0.5%, 0.75%, 0.1%, 0.15%, 0.2% respectively. After reaching 2%, it is increased by 1% per level.

Table 1 Material properties of steel

Material	E(MPa)	σ_y(MPa)	ε_y	σ_u(MPa)	ε_u
Steel	206000	287	0.00175	435	0.163

3 Parameter Analysis

3.1 Failure Mode

Figure 3 shows the failure modes of specimens, which are divided into two categories: the shear failure of the joint core area and the plastic hinge failure of the beam end. When $B/b_f>1; t_d/t_f>1.4$, the failure mode changes from the core failure of the joint to the plastic hinge failure at the beam end.

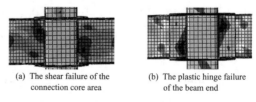

(a) The shear failure of the connection core area   (b) The plastic hinge failure of the beam end

Figure 3 Failure modes

3.2 Hysteretic Curves

Figure 4～Figure 5 show the hysteretic curves of two groups of specimens. The hysteretic curves are full "fusiformis", which shows that the joints have excellent seismic performance and energy consumption capacity. The bearing capacity of the specimens with the shear failure of the joint core area is relative undesirable. The bearing capacity increases continuously without a descending stage because only slight shear deformation occurs in the joint core area at the later stage of loading. The ultimate bearing capacity of the specimens with plastic hinge failure at the end of the beam increases rapidly with loading. At the later stage of loading, the hysteretic curve shows a significant decline because of the weakening of the bearing capacity at the end of the beam, and the ultimate bearing capacity is greater than that of the specimens with joint core failure. The hysteretic curves of the joints become fuller with the increasing of the value of each parameter.

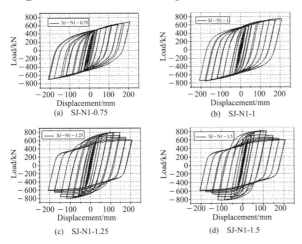

Figure 4　Hysteretic curves of SJ−N1 series specimens

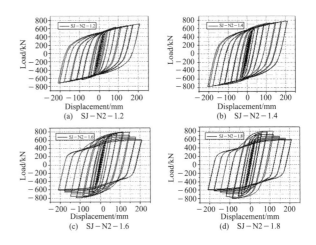

Figure 5　Hysteretic curve of SJ-N2 series specimens

3.3 Skeleton curve and bearing capacity

Figure 6 presents skeleton curves comparison diagram of two groups of specimens. It shows that the two groups have three stress stages: elastic stage, elastic-plastic stage and failure stage. At elastic stage and elastic-plastic stage, the shape and variation trend of skeleton curves are approximately identical. The skeleton curves of the two failure modes show great difference at the failure stage. The skeleton curves of the specimens with core failure always show an increasing trend without a descending stage. The other group of skeleton curves has a descending stage, and the displacement increases sharply with smaller load changes, which means the specimens have good deformability. According to the bearing capacity of the joint, the ultimate bearing capacity of the joint shows an increasing trend with the increase of the value of each parameter. Due to the appearance of plastic hinges, the bearing capacity of specimens with plastic hinges failure of the beam end will decrease when they are at the failure stage. The parameters B/b_f and t_d/t_f have great influence on the ultimate bearing capacity, and the ultimate bearing capacity increases obviously with the increase of parameters' value.

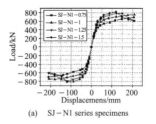
(a) SJ-N1 series specimens

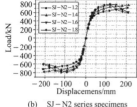
(b) SJ-N2 series specimens

Figure 6 Skeleton curves

3.4 Stiffness Degradation

Figure 7 presents the stiffness degradation curves of the two groups of specimens, and the stiffness exhibits a similar degradation trend. From the figures (a) and (b), it shows that the curves of SJ-N1 and SJ-N2 series specimens are relatively scattered during the whole loading process. The stiffness of the specimens increases with the increase of the values of B/b_f and t_d/t_f.

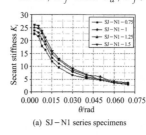
(a) SJ-N1 series specimens

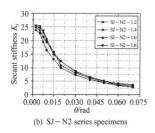
(b) SJ-N2 series specimens

Figure 7 Stiffness degradation curves

3.5 Ductility Performance

Table 2 shows the value of ductility performance of two groups of specimens. The ductility coefficients of the specimens are between 2.5 and 3.7, and the joints have great ductility performance. In practical engineering, the brittle failure of weld seam may lead to the ductility degradation of the specimens with joint core failure.

Table 2 Displacement ductility coefficient

Specimen number	P_y kN	Δ_y mm	P_u kN	Δ_u mm	Ductility coefficient
SJ-N1-0.75	503.14	41.22	693.48	142.85	3.466
SJ-N1-1	579.66	41.98	753.23	146.99	3.501
SJ-N1-1.25	644.22	42.36	797.75	133.85	3.160
SJ-N1-1.5	679.15	41.21	820.39	118.49	2.875
SJ-N2-1.2	534.92	40.03	716.30	148.84	3.718
SJ-N2-1.4	611.86	41.37	778.94	146.23	3.535
SJ-N2-1.6	644.22	42.36	797.75	133.85	3.160
SJ-N2-1.8	661.02	40.42	803.75	104.02	2.573

4 Conclusions

(1) The type hysteretic curve of this type of joint is full "fusiformis", and the ductility coefficient of this type of joint is between 2.5 and 3.7, which has good ductility and seismic performance.

(2) Two parameters of the reinforcing ring have great influence on the failure mode of the joint, hysteretic performance, skeleton curve, ultimate bearing capacity and stiffness degradation of the joints. With the increasing of the two parameters, the failure mode of the joint varies from the failure of the core area of the joint to the failure of the plastic hinge at the end of the beam, and the hysteretic curves of the joints become fuller. The skeleton curve of the joints in the elastic-plastic stage increases significantly, and the ultimate bearing capacity and stiffness of the joints also increase obviously.

(3) It is suggested that the ratio of the length of the external reinforcing ring along the beam side to the width of the beam flange should be greater than 1 and

the ratio of the thickness of the external reinforcing ring to the thickness of the beam flange should be greater than 1.4 at design stage, which ensures that the joints not only have good seismic performance, but also can avoid the shear failure of the joint core area.

References

[1] Zhang Chunlei, Han Junke. The seismic behavior of cold-formed rectangular tube-steel column and H-shaped beam connections [J]. Industrial Building, 2009, 39 (S1): 466-468+474. (in Chinese)

[2] Yang Xiaojie, Zhang Long, Li Guoqiang. Hysteretic performance of end plate connections between rectangular hollow section columns and H-shaped beams using through-bolts [J/OL]. Progress in Steel Building Structures, 2013, 15(04): 16-23. (in Chinese)

[3] Okamoto T, Maeno T, Hisatoku T, et al. A design and construction practice of rectangular steel tube columns infilled with ultra-high strength concrete cast by centrifugal force [C]. Proceedings of the Third International Conference on Steel-Concrete Composite Structure, 1991: American.

[4] Usami, T. and Ge, H. B. Ductility of concrete-filled steel box columns under cyclic loading. Journal of Structural Engineering. 1994: American.

[5] China Association for Engineering Construction Standardization. GB 50017-2003 Code for design of steel structures [S]. Bei Jing: China Planning Press, 2003. (in Chinese)

[6] Japan Architecture Association. Construction pointer and explanation of steel pipe structure design[S]. 2002: Japan.

[7] China institute of building standard design & research. Detailed diagram of connection structure of steel structure for multi and high rise civil buildings [S]. Beijing: China Planning Press, 2016. (in Chinese)

[8] Li Xingrong, Wei Angcai, Qin Bin. Design Manual of connection of steel structure[M]. Beijing: China Building Industry Press, 2014. (in Chinese)

[9] Tongji University, Hang xiao steel structure. CECS 159-2004 Technical specification for rectangular concrete-filled steel structure[S]. Beijing: China Planning Press, 2004. (in Chinese)

[10] AISC, Seismic Provisions for Structural Steel Buildings. American Institute of Steel Construction, INC, Jun 22nd, 2010: American.

(10) Study on Seismic Behavior of Steel Structure Joints with Inner Diaphragms and Outer Ring Plates

Ju Jiachang, Xu Weixiao, Yang Weisong, Yu Dehu

School of Civil Engineering, Qingdao University of Technology Yu Fengbo

Qing Dao, Hehai Green Building Innovative Technology Company

Abstract

Beam-to-column joint is the key part of steel frame. Its mechanical behavior is complex and the connection performance directly affects the overall mechanical performance of the whole structure. In this paper, two kinds of joints, inner diaphragm and outer ring plate are studied. Nonlinear finite element analysis of both joints is carried out by ABAQUS software. Hysteresis curve are obtained under low cyclic loading. The stress distribution, ultimate load, ductility and energy dissipation of the joints are studied by simulation results.

Key Words

Steel Structure Joints, Inner Diaphragms, Outer Ring Plates

1 Introduction

"Code for seismic design of buildings" (GB 50011 − 2010)[1] clearly requires that "strong members and stronger joints" should be followed in the design process. According to the relevant experimental studies, the energy dissipation capacity of joints accounts for about 30% of the energy dissipation of the whole frame[2]. So we choose nodes for research. Compared with the traditional inner diaphragm joints, the proposed outer ring plate joints have the following advantages: (1) solving the problem of difficult processing of the inner diaphragm in the steel tube when the existing steel tube column is connected with the steel beam; (2) solving the problem of difficult concrete placement in the steel tube caused by adding the inner diaphragm; (3) to provide convenience for steel structure construction and save comprehensive cost.

2 Model Introduction

The two kinds of beam-to-column joints have the same size: the column height is 3m, the beam length is 1.5m, the column is box steel, and the beam is I-beam. The outer ring plate joints have two outer ring plates, and the inner diaphragm has two inner diaphragms. Detailed drawings of the two kinds of joints are shown in Figure 1. The section size of the column is 450 mm × 200 mm ×

12 mm×12mm, the cross section dimension of the beam is H350 mm×180 mm ×10 mm×12 mm, and the material is Q345B.

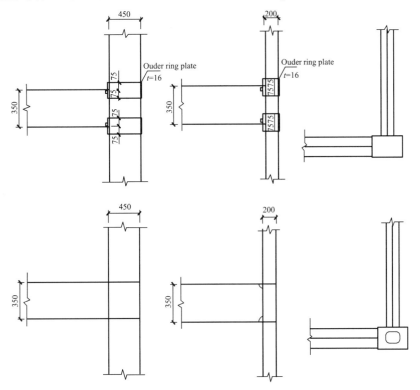

Figure 1 Details of outer ring plate and inner partition joint

3 Finite Element Analysis

Both joints are modeled by solid elements. The constitutive relationship of Q345 steel is modeled by the three broken line following enhancement model as shown in Figure 2[3]. The elastic modulus is 200 000 MPa, Poisson's ratio is 0.3, yield strength is 360 MPa and tensile strength is 530 MPa. The grids of the two kinds of nodes are all divided by 25 mm, and different grids are processed in special parts. The outer ring plate is swept and refined at the opening of the beam, and the flange of the beam increases the number of grids; the inner diaphragm node sweeps the inner diaphragm, and then the grids are refined around the elliptical buckle hole.

The simulation is carried out without

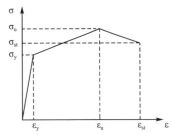

Figure 2 Three broken lines of steel composition

considering the weld strength, assuming that the weld is strong enough to avoid damage at the weld. The restraint mode is hinge joint. The axial compression ratio is 0.3 at the end of the column. The calculated force is applied to the end of the column in the form of pressure, and low cyclic loading is applied to the end of the beam.

The loading system is loaded according to the displacement angle between the layers. The displacement angle between the layers is the ratio of the displacement of the beam end to the distance from the loading point to the center of the column. When loading, the displacement is applied to the end of the beam. The loading history is shown in Table 1. The loading is completed according to Table 1. After that, 0.01 rad is added each time after loading until destruction. Each cycle is loaded only once.

Table 1 Loading program

Load level	Displacement amplitude	Cycle times	Interlar drift angle
1	±5.63	1	0.00375
2	±7.5	1	0.005
3	±11.5	1	0.075
4	±15	1	0.01
5	±22.5	1	0.015
6	±30	1	0.02
7	±45	1	0.03
8	±60	1	0.04
9	±75	1	0.05
10	±82.5	1	0.055
11	±90	1	0.060

After loading, the stress nephogram of the outer ring plate at the end of loading is shown in Figure 3, and the stress nephogram of the inner partition joint is shown in Figure 4.

Figure 3 Stress nephogram of the outer ring plate

Under cyclic loading, the relationship between structural resistance and deformation is called "hysteresis curve". The hysteresis curve of a structure or member can be expressed as the relationship between bending moment and rotation angle, load and displacement, or stress and strain[4]. The hysteresis curves of the two nodes are shown in Figure 5~6, and the skeleton curves are compared with those in Figure 7.

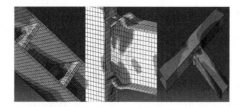

Figure 4 Stress nephogram of the inner partition joint

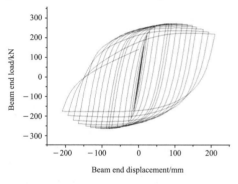

Figure 5 Hysteresis curves of the outer ring plate

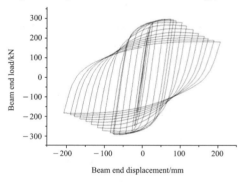

Figure 6 Hysteresis curves of the inner partition joint

Ductility refers to the deformation capacity of a structure or member under certain load conditions after yielding or before failure. It is an important index to evaluate the seismic performance of the structure or member, and reflects the deformation capacity of the structure,

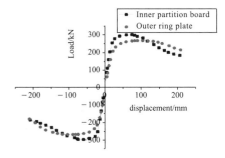

Figure 7 Skeleton curves

member or material in the non-elastic stage. The larger the ductility ratio is, the stronger the ability of identifying mechanism to dissipate seismic energy is, and the better the seismic performance is. Ductility can be expressed by load-deformation curve or ductility coefficient. Ductility coefficient refers to the ability of structure to continue to increase deformation after yielding. The displacement ductility coefficient is the ratio of the displacement to the yield displacement when the node is subjected to the ultimate failure[5]. The equivalent yield displacement calculation is simplified as Figure 8.

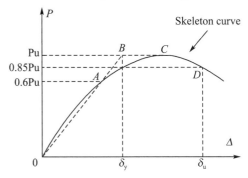

Figure 8 Sketch of equivalent yield displacement calculation

Energy dissipation capacity is the ability to absorb energy dissipation due to deformation of structural members under earthquake. The energy dissipation capacity index of the structure is used to characterize the energy dissipation coefficient C_e. The larger the C_e is, the more the energy consumption of the specimen is. When the force is large, the formula is as following: $C_e = S_{\widehat{ABCD}}/(|S_{\triangle OBE}| + |S_{\triangle ODF}|)$, $S_{\widehat{ABCD}}$ tantalum is the area of the hysteretic loop under cyclic loading, and $S_{\triangle OBE} + S_{\triangle ODF}$ are the sum of the areas of the triangle consisting of the vertical line from the positive and negative maximum load point to the axis [6], as shown in Figure 9.

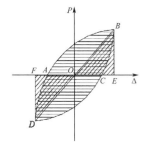

Figure 9 Calculation of energy dissipation coefficient

The ultimate load, ductility and energy dissipation of the joints can be obtained by analyzing the hysteretic curves obtained from the simulation as shown in Table 2.

Table 2 Performance index

Node name	Ultimate bearing capacity	Ductility coefficient	Energy dissipation coefficient
Inner partition node	298 kN	5.7	2.7
Outer ring plate joint	266 kN	6.5	2.4

4 Conclusion

The outer ring plate joints have almost the same load bearing capacity, energy dissipation and ductility compared with the traditional internal diaphragms. But it has the advantages of convenient construction and convenient placement of concrete, and is more convenient for engineering application.

Acknowledgments

The research is supported by the National Natural Science Foundation of China (Project Nos. 51608287, 51508295), the Key Research and Development Project of Shandong Province (Project No. 2018GSF120004).

References

[1] Code for seismic design of buildings GB 50011—2010 [M]. China Architecture & Building Press, 2010.

[2] Zhang Shanfeng. Analysis on rigid connection structure of H type box column. [J] Journal of Steel Structure, 2005, 6 (28): 68-71.

[3] Yu Yousheng, Zhang Yanyan, Li Jianfeng, Wang Yan. Hysteretic behavior of a new type of beam-column assembling rigid joint[J]. Progress in Building Steel Structure, 2014, 16 (02): 1-5+12.

[4] Ma Hui. Finite element analysis of flange-enlarged and flange-strengthened joints of steel frame beams [D]. Qingdao University of Technology, 2010.

[5] Gao Yang. Experimental study on hysteretic behavior of K type gap square pipe joints [D]. Harbin Institute of Technology. 2008.

[6] Li Zeshen, Li Xiumei, Zheng Xiaowei and Zhang Keshi. Experimental study and numerical analysis of hysteretic behavior of semi-rigid T-shaped steel beam-column connections [J]. Journal of Architectural Structures, 2014, 35 (07): 61-68.

(11) The Characteristics of Underground Comprehensive Pipe Gallery Structure and Waterproof Practices

Han Jinpeng, Liu Junwei, Zhao Yanping, Huang Xiaoyi

School of Civil Engineering, Qingdao University of Technology

Abstract

Municipal pipelines are like the blood vessels of a city, while the underground pipe gallery is like the protective layer of municipal pipes, which greatly facilitates the management and maintenance of municipal pipelines. This paper first expounds the development process and current situation of underground comprehensive pipe gallery at home and abroad, and then summarizes the following four structures of precast pipe gallery in China based on the concept of smart city in China: Precast steel corrugated pipe gallery, semi-precast concrete pipe gallery, sectional precast concrete pipe gallery and superimposed precast concrete pipe gallery, to provide reference for the design of precast comprehensive pipe gallery which can stand the test of engineering and practice.

Key Words

Underground Comprehensive Pipe Gallery

1 Introduction

As a major part of urban infrastructure construction, municipal pipelines have become the basis and guarantee for urban survival and development[1]. In the past, most urban municipal pipelines in China were directly buried under roads, which lacked the idea of underground management, resulting in congestion and chaos in the shallow underground space of urban roads[2]. Therefore, it is urgent to strengthen the management and protection of municipal pipelines. The underground comprehensive pipe gallery is a modern and intensive urban infrastructure formed by placing two or more urban pipelines in the same artificial space[3]. After 2014, the government issued a series of preferential policies and guiding opinions on the planning, capital and technology of the construction of the comprehensive pipe gallery, and encouraged the large-scale construction of the comprehensive pipe gallery in cities. When the underground comprehensive pipe gallery is designed and built, 50-100 years of development and expansion space should be reserved according to the urban development plan[4], this requires the comprehensive pipe gallery design must be strict and optimal, and achieve fine construction[5].

2 Domestic and Foreign Construction Statuses

2.1 Foreign Construction Status At Abroad

In France as early as 1833, a prototype of the underground comprehensive pipe gallery appeared. The construction of the underground pipe gallery began in London in 1861; In 1893, the underground pipe gallery built in Germany added gas pipelines on the basis of keeping the electric power, communication, water pipelines. Then came the construction of underground pipe gallery in Spain, Sweden, the United States, Canada, Russia and Japan. Spain is the first country to systematically plan the comprehensive pipe gallery. The underground pipe gallery in the United States promoted the waterproof research of it greatly; Although Japan started late, its underground pipe gallery has been developing rapidly. In less than 80 years, Japan has become a country with the fastest construction speed, the most complete planning, the most perfect regulations and the most advanced technology in the world. It is worth mentioning that the underground pipe gallery planning network in Madrid adopts the screen network layout, which is novel and practical.

2.2 Domestic Construction Status

China's underground pipe gallery started late. In 1958, China built its first underground pipe gallery under Beijing's Tian'anmen Square; In 1977, an underground pipe gallery with a length of 500 meters was built during the construction of "Chairman Mao Zedong Memorial Hall"; In 1991, Taipei's first underground pipe gallery was put into operation; In March 2003, Beijing Zhongguancun built a 1.9km underground pipe gallery in three years, which was one of the relatively complete projects in the initial construction stage of underground pipe gallery in China.

In recent years, with the continuous improvement of the construction environment of urban comprehensive pipe gallery and the increasing investment of capital, the underground pipe gallery in China has been constructed like mushrooms after rain.

3 The Structure of Existing Precast Pipe Gallery

As a traditional construction method, cast-in-place construction gradually reveals its shortcomings in the construction of urban pipe gallery, which makes us have to study and improve the construction of pipe gallery[6]. Compared with the traditional cast-in-place pipe gallery, the precast underground pipe gallery has obvious advantages in shortening construction period, saving materials, saving energy and environmental protection[7,8]. Precast comprehensive pipe gallery has good economic, social and environmental benefits and has been widely used in Japan, the United States and other countries[9]. In contrast, the construction of the precast pipe gallery in China has just started. At present, several kinds of precast comprehensive pipe gallery in China are mainly precast steel corrugated pipe gallery, semi-precast concrete pipe gallery, sectional precast concrete pipe gallery and superimposed precast concrete pipe gallery.

3.1 Precast Steel Corrugated Pipe Gallery

Precast steel corrugated pipe gallery refers to the combination of multiple steel corrugated pipes to form a structural form of the whole pipe gallery. Precast steel corrugated pipe gallery (Figure 1) mainly has the advantages of simple construction, short period, strong anti-deformation ability, durability, stability and reliability, economic and environmental protection as well as rapid mass production.

3.2 semi-precast concrete pipegallery

Semi-precast concrete pipe gallery (Fig-

Figure 1 Precast steel corrugated pipe gallery

ure 2) refers to that after the precast components of the underground comprehensive pipe gallery are installed in place, cast-in-place concrete is used to connect each component at the joint, so that the components form a whole. Compared with cast-in-place concrete pipe gallery, semi-precast concrete pipe gallery has the advantages of rapid construction, low technical requirements for construction personnel and relatively economic. Compared with the full-precast concrete pipe gallery, it also has the advantages of good waterproof performance and integrity.

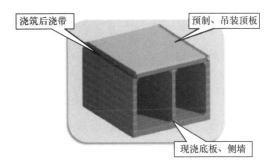

Figure 2 Semi-precast concrete pipe gallery

3.3 Sectional Precast ConcretePipe Gallery

Sectional precast concrete pipe gallery adopts factory precast components and field assembly. The advantage of the sectional precast concrete pipe gallery is that the construction speed is fast, and a series of works such as the reinforcement and formwork of the traditional cast-in-place reinforced concrete and the concrete pouring and curing are completed in the precast factory, which greatly reduces the time of site construction. Different from outdoor work, the steel bar is tied in place, the protective layer thickness is even, the concrete vibrating is compacted, and the maintenance is sufficient, so that the finished product quality of the comprehensive pipe gallery can be guaranteed to the maximum extent[10] (Figure 3, Figure 4).

Figure 4　Single compartment full-precast concrete pipe gallery

Figure 5　Double compartments full-precast concrete pipe gallery

Figure 3　Sectional precast concrete pipe gallery

At present, the common sectional precast concrete pipe gallery is divided into three types according to different requirements for pipeline layout: single compartment, double compartments and upper and lower combined compartments (Figure 5, Figure 6).

3.4　Superimposed Precast Concrete Pipe Gallery

The superimposition assembly technology is mostly used in the underground multi-storey multi-span comprehensive

Figure 6　Upper and lower combined compartments full-precast concrete pipe gallery

pipe gallery. The precast floor, side wall and roof structure are overlapped, and concrete is poured on the joint and waterproof measures are taken to form the comprehensive pipe gallery. The superposed precast concrete pipe gallery has the advantages of good mechanical performance, structural safety and low construction cost (Figure 7).

4　Conclusion

It is urgent to develop the assembly structure and waterproof technology more suitable for the underground pipe gallery,

Figure 7 Schematic diagram of superimposed precast concrete pipe gallery

to create a smart city, and to improve the living environment, with the continuous advancement of innovation, quality and green concepts in related research and development, although the current construction process is still dominated by cast-in-place, precast assembly technology is obviously the direction of future development of comprehensive pipe gallery. This paper expounds the origin, development, classification and characteristics of urban underground pipe gallery at home and abroad, summarizes and analyzes the structure of existing underground pipe gallery, provides ideas for the design and construction of underground pipe gallery for relevant practitioners, and promotes the construction of new smart city.

References

[1] Gui Xiaoqin. 2011. The Incentive Mechanism for Financing of the Underground Comprehensive Pipe Gallery Construction[J]. Chinese Journal of Underground Space and Engineering.

[2] Yang Chao. 2016. Research on the Current Situation, Problems and Countermeasures of Comprehensive Pipe Gallery in China [J]. Technology and Market.

[3] Qian Qihu, Chen Xiaoqiang. 2007. Research on the Current Situation, Problems and Countermeasures of Underground Comprehensive Pipe Gallery Development in China and Abroad [J]. Chinese Journal of Underground Space and Engineering.

[4] Yu Changjun. 2016. Brief Analysis of Underground Comprehensive Pipe Gallery [J]. Highway.

[5] Zhou Jianmin. 2016. Design Form and Applicability Analysis for Deformation Seam and Joint of Comprehensive Pipe Gallery[J]. Special structure.

[6] Shi Liguo. 2017. Construction Technology of Two-piece Unit Precast Pipe Gallery[J]. Construction technique.

[7] Xue Weichen, Hu Xiang. 2009. Experimental Research on Mechanical Properties of Precast Prestress Comprehensive Pipe Gallery in Shanghai Expo Area[J]. Special structure.

[8] Wang Jun. 2016. Review and Thinking on Development of Building Industrialization in China [J]. China Civil Engineering Journal.

[9] Huang Yingchang. 2003. Elastic Sealant and Adhesive[M]. Chemical Industry Press.

[10] Su Yanyu. 2017. Application of Precast Technology in Comprehensive Pipe Gallery[J]. Sichuan Cement.

(12) Summary of Research on Pile-soil Dynamic Response under Lateral Cyclic Loading

Huang Xiaoyi, Wang Mingming, Liu Junwei

School of Civil Engineering, Qingdao University of Technology

Abstract

The offshore pile foundation has attracted extensive attention because of the development of offshore wind power in recent years. And the researchers conducted the studies on the pile-soil dynamic response mechanism under cyclic loading, through field experiments, numerical simulation and theoretical analysis. However, the study on bearing characteristics of pile foundation under lateral cyclic loading lags behind the development of engineering practice. By summarizing the bearing characteristics of pile foundation under cyclic loading at home and abroad, a two-dimensional particle flow program is used to simulate the action process of model piles under cyclic loading. The macro and micro mechanism of model piles under lateral cyclic loading is studied by simulating different pile diameters and loading modes.

Key Words

Pile-Soil Dynamics, Lateral Cyclic Loading

1 Introduction

With the rapid development of the economy, China's energy demand has been greatly increased. Additionally, China has extensive coastline, which is consistent with global climate agreements, and offshore wind power is growing increasingly. In actual design application, the effect of lateral cyclic loading on pile foundation is a non-negligible factor. Single pile foundation of marine fan mainly involves wind load, wave load and other cyclic loading with significant periodicity. Because of the uncertainty of cyclic loading, pile-soil response mechanism under cyclic loading is more complicated than response mechanism under static load. Under repeated lateral cyclic loading, the foundation stiffness varies with the change of load cycles, which may lead to the changes in the natural frequency of the system, and is likely to produce unplanned system resonance, as well as other consequences, such as excessive tile, costly repair or even complete shutdown. Therefore, it is essential to understand the long-term response of the fan base so that a method for predicting frequency and long-term changes can be established.

2 Progress of Model Pile Test under Cyclic Loading

As early as in 1972, Broms[1] conducted field full-scale test on precast concrete square piles under different cyclic loading amplitudes. It is found in the study that, when the amplitude of cyclic loading applied to the pile tip exceeds a critical value, the cumulative deformation of the pile tip is mutated. Then, Matlock & Holmquist[2] studied the bearing characteristics of pile foundation in the soft clay through cyclic loading of displacement control in 1976. It is found in the study that, when the cyclic loading is greater than or equal to 3/4 of bearing capacity of pile foundation, the pile foundation will appear continuous accumulative deflection. In 1979, Poulos[3] studied ultimate bearing capacity of single pile under cyclic loading and load-stiffness response mechanism, and predicted the influence of water pressure of lateral porosity on elasticity modulus of soil around pile and lateral friction through effective stress principle. In 1980, Chan & Hanna[4] conducted model pile test with medium and small scale of dry sand under cyclic loading. It is found in the study that, under the cyclic loading, pile shaft stress will be redistributed and accumulative settlement of pile tip increases with the increase of number of cycles. In 1981, Kraft et al.[5] conducted field test of steel pipe pile, where, the pile length was 15 m, and pile diameter was 356 mm, the test pile was penetrated into different depths with the casing to study the cyclic weakening characteristics of pile foundation at different depths. In 1991, Lee & Poulos[6] conducted cyclic loading test of small scale on the friction pile in the sand, and found that, permanent cumulative settlement under bilateral cyclic loading was more serious than that of unilateral cycling. In 2010, Leblanc et al.[7] conducted a series of experiments and tests on the rigid pile under long-term cyclic loading, and concluded that, the action of cyclic loading can always increase the secant stiffness of pile shaft, and unrelated to relative density. In the same year, Li[8] studied the influence of axial performance of single pile under cyclic loading through the centrifugal model of single pile under cyclic loading. It is found in the study that, with the increase of cyclic loading amplitude, the accumulative settlement of pile tip increases constantly, and the pile sinking mode has a greater impact on the vertical bearing capacity of pile shaft. In 2012, Wichtmann et al.[9] studied dynamic properties of clay soil through resonant column test, and concluded that main factors that affect the weakening and deformation of soft clay include initial shear stress, rotation extent of primary stress, shear frequency and over-consolidation ratio of soil, etc.

Although the study on pile-soil dynamic response under cyclic loading started late in China, domestic scholars have conducted a lot of valuable researches, and accumulated more research achievements. In 2000, Peng Xiongzhi et al.[10] carried out in situ and indoor model test through vertical vibration characteristics of pile foundation under cyclic loading, and found that, under dynamic load, the pile lateral friction decreases parabolically along the depth, and the decrease amplitude decreases with the increase of the depth. In 2005, Zhang Ga[11] set up an elastic-plastic damage constitutive ("EPDI") model based on the tests, in order to describe main mechanical properties of coarse grained soil and structure interface. The model can uniformly describe the strain-softening of contact surface, shear dilation rules, anisotropy and other main mechanical properties under single or cyclic loading conditions. In 2009, Huang Yu et al.[12] conducted indoor model pile test and found that, cyclic loading ratio and number of cycles are main factors that affect the weakening and cumulative settlement deformation of pile-soil interface. In 2011, Chen Renpeng et al.[13] conducted model pile test of large scale with large-scale indoor model box, and studied bearing capacity and deformation characteristics of rigid pile under cyclic loading systematically, concluding that, cyclic loading ratio has significant influence on the accumulative settlement and lateral stress development of pile foundation. In 2014, Huang Maosong et al.[14] proposed elastic-plastic damage weakening model of soft clay under attenuation of cyclic loading, and studied the influence of lateral cyclic loading, number of cycles, and soil stiffness on weakening characteristics of vertical bearing capacity of single pile.

3 Establishment of Numerical Simulation Model

The DEM model used in this paper is 2 400 mm × 2 400 mm (length × height), and the outermost four walls are generated counterclockwise to simulate the model caisson, and numbered to ensure the inner wall of the model caisson is the effective side. The particle model[15] generated by GM method is as shown in Figure 1.

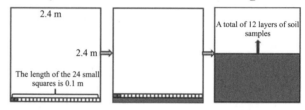

Figure 1 Preparation of soil samples by GM method

By GM method, the soil sample model is divided into a number of small grids, and particles are generated from left to right and from bottom to top, effectively avoiding compaction in the formation process of soil layers. The maximum particle size of the model is 3.52 mm, the minimum particle size is 2.25 mm, the median particle diameter D_{50} = 2.92 mm, the coefficient of nonuniformity $Cu = D_{60}/D_{10} = 1.26$, and the particle grading curve is as shown in Figure 2.

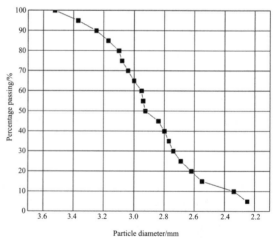

Figure 2 Particle grading curve of soil sample

Based on the pile generation method proposed by Duan[16], the model pile is generated, comprised of particles at a radius of 1.125 mm. The particles overlap each other, and the distance between the centers of two adjacent particles is d_{pp} (0.2R), as shown in Figure 3. The diameter of the particles forming the pile is much smaller than the pile shaft diameter, the distance between particles is short, and the particle surface is smooth, with its roughness close to the initial set value, the contact force between the particles and the pile is in the same direction as the axial direction of the pile, so that the calculation of the axial resistance of pile is easier and more accurate.

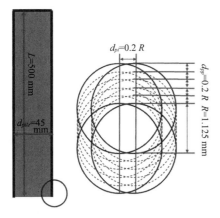

Figure 3 Pile formation

4 Conclusion

In summary, the scholars at home and abroad did a lot of researches on dynamic analysis for bearing characteristics of pile foundation under cyclic loading, and had made great progress. By using the foregoing simulation, it can further explain the action process of cyclic loading from the mesoscopic level. Additionally, by connecting macromechanical behavior of pile shaft in loading process and meso-structure change of soil around pile after being influenced by cyclic loading, it has improved the in-depth understanding of pile foundation under cyclic loading.

(1) This paper analyzes the evolution law of pile tip resistance, the outer unit frictional resistance and the inner unit frictional resistance of the pile under lateral cyclic loading, to reveal the influence mode of different pile diameters on soil plug, displacement of soil around pile and pile driving resistance, presenting a research on the lateral cyclic bearing characteristics of open-ended pipe pile at macroscopic and mesoscopic levels.

(2) The paper studies the pile displacement under different cyclic loadings, displacement of soil around pile, pile tip load displacement relations, distribution law of pile lateral friction and lateral pressure, and analyzes the influence of loading mode, velocity and pile diameter on the model pile.

References

[1] Broms BB. Bearing capacity of cyclically loaded piles. Swedish Geotechnical Institute, 1972, No 44, pp. 1-16.

[2] Matlock H, Holmquist DV. A model study of axially loaded piles in soft clay. Report to the American Petroleum Institute. The University of Texas, Austin, 1976.

[3] Poulos HG. Development of an analysis for cyclic axial loading of piles [C]. Proc. 3rd Int. Conf. Num. Meth. Geomechs, 1979, 4:1513-1530.

[4] Chan SF, Hanna TH. Repeated loading on single piles in sand[J]. Journal of the Geotechnical Engineering Division, ASCE, 1980, 106(2):171-188.

[5] Kraft LM, Cox WR, Verner EA. Pile load tests: Cyclic loads and varying load rates[J]. Journal of the Geotechnical Engineering Division,

1981, 107(GT1): 1-12.
[6] Lee CY, Poulos HG. Tests on model instrumented grouted piles in offshore calcareous soil[J]. Journal of Geotechnical Engineering, 1991, 117(11):1738-1753.
[7] Leblanc C, Houlsby GT, Byrne BW. Response of stiff piles in sand to long-term cyclic lateral loading. Géotechnique, 2010, 60(2): 79-90.
[8] Li Z, Haigh SK, Bolton MD. The behavior of a single pile under cyclic axial loads. Geotechnical Special Publication, 2010: 143-148.
[9] Wichtmann T, Triantafyllidis T. Behavior of Granular Soils under Environmentally Induced Cyclic Loads. Mechanical Behavior of Soils under Environmentally Induced Cyclic Loads. Springer, Vienna, 2012: 1-136.
[10] Peng XZ, Zhao SR, Luo SX, Wang AL. Dynamic model tests on pile foundation of high-speed railway bridge. Chinese Journal of Geotechnical Engineering, 2002, 24(2): 218-221.
[11] Zhang G, Zhang JM. Unified modeling of soil-structure interface and its test confirmation. Chinese Journal of Geotechnical Engineering, 2005, 27(10):1175-1179.
[12] Huang Y, Bai J, Zhou GM, Huang Q. Model tests on settlement of a single pile in saturated sand under unilateral cyclic loading. Chinese Journal of Geotechnical Engineering, 2009, 31(9): 1440-1444.
[13] Cheng RP, Ren Y, Chen YM. Experimental investigation on a single stiff pile subjected to long-term axial cyclic loading. Chinese Journal of Geotechnical Engineering, 2011, 33(12):1926-1933.
[14] Huang MS, Liu Y. Numerical analysis of axial cyclic degradation of a single pile in saturated soft soil based on nonlinear kinematic hardening constitutive model. Chinese Journal of Geotechnical Engineering, 2014, 36(12):2170-2178.
[15] Duan N, Cheng Y. A modified method of generating specimens for a 2D DEM centrifuge model. Geo-Chicago, American Society of Civil Engineers, 2016: 610-620.
[16] Duan N. Mechanical Characteristics of Monopile Foundation in Sand for Offshore Wind Turbine. UCL (University College London), 2016.

(13) Preliminary Study on Laboratory Model Test under Vertical Cyclic Loading

ZhaoYanping, Wang Mingming, Liu Junwei

School of Civil Engineering, Qingdao University of Technology

Abstract

The dynamic response of "soil plug-pipe pile- pile side soil" system is the key factor to determine the safety of infrastructure operation. The formation of soil plug makes the pile driving resistance of open pipe pile different from that of closed pipe pile, which not only includes c of pile, resistance of pile tip, but also inner friction resistance of pile. This paper expounds the current situation at home and abroad from three aspects, including test method, test scheme and test results. In view of the existing deficiencies in the current research, a large-scale cyclic loading test device is designed, which has important theoretical value and practical engineering significance to evaluate the bearing characteristics of pipe pile under long-term cyclic load.

Key Words

Cyclic Loading, Model Test

1 Introduction

Pipe piles have become the most important form of foundation in China. Their application range not only covers traditional industrial and civil buildings, but also extends to highway, railway, energy, port and other engineering construction fields. In particular, with the rapid construction of major infrastructure of "Two Districts, One Circle and One Belt" in our province (such as Jinan-Qingdao High-speed Railway, Dongjiakou port and Qing Rong Intercity Railway), pipe piles can be applied in a wider range. In these projects, the pile foundation not only bears the static load generated by the superstructure dead weight during its service life, but also bears the vertical cyclic load caused by wind, wave or vehicle, which has obvious periodicity. Under cyclic loading, the bearing characteristics of pile foundation are totally different from the performance under static load. There are many problems such as weakening of bearing capacity and cumulative settlement, especially for those structures which are sensitive to the foundation uneven settlement such as high-speed traffic and wind generator. The cumulative settlement of pile foundation under cyclic load has become the most important parameter to control in design. However, as the interaction mechanism between pile and soil under cyclic load is very complicated and influenced by many factors, such as cycle index, loading mode, stress path, etc., it is difficult for existing theories to accurately reflect the actual working situation between pile and soil. Therefore, urgent requirements are put forward on how to consider the design of pile foundation under cyclic load.

Paik and Lee(1993)[1] sunk open-ended pipe pile into different sands to study the effect of different soil conditions on load transfer mechanism of open-ended pipe pile. It is measured that the bearing capacity of the open-ended pipe pile includes lateral friction resistance, pile tip resistance and inner friction resistance. Different soil conditions have some influence on the load capacity of three parts, and a one-dimensional analysis method for calculating the soil plug pressure coefficient is put forward.

Al-Douri & Poulos (1995)[2,3] etc. carried out experimental studies. It is found that cyclic load has significant influence on the mechanical properties of pile-soil interface.[4,5]

Peng Xiongzhi (2000)[6] etc. carried out model pile test. It is found that the cyclic load ratio and cycle times are the main factors influencing the weakening of the pile-soil interface and the cumulative settlement.

Huang Maosong(2013,2014)[7] etc. put forward that the elastic and plastic damage reduction model of soft clay with cyclic attenuation effect can be studied from the influence modes of cyclic load level, cycle times, soil stiffness index on the weakening characteristics of vertical bearing capacity of single pile.

Zhang Ga(2005)[8] etc. studied the cyclic shear characteristics of the interface between soil and structure (pile) by using the interface cyclic loading shear apparatus.

Bai Shunguo etc.[9] conducted relevant model tests on the cyclic bearing characteristics of soil-cement pile composite foundation under cyclic loading.

Fan Qinglai etc.[10] conducted relevant research on load deformation characteristics under cyclic load of bucket foundation and bearing capacity of foundation under composite loading.

Yang Longcai etc.[11] studied the bearing capacity and deformation characteristics of high-speed railway bridge pile foundation in soft clay under long-term axial cyclic load through field test.

Huang Yu etc. carried out relevant model experiment study on the cumulative deformation characteristics of a single pile under axial cyclic loading in saturated sandy soil.

To sum up, domestic and foreign researches on long-term weakening of single pile under single-degree-of-freedom cyclic loading (vertical or horizontal) are abundant. However, most of them are developed for closed pipe piles or cast-in-place piles, without involving the weakening process of soil plug and its influence on the circulation characteristics of open-ended pipe pile. There are many researches on the existing double-wall model pipe piles in foreign countries, but few in China. However, the double-wall model pipe pile studied abroad is small in size and has obvious size effect. Secondly, the influence of the form of pile tip on the penetration of pile is not taken into account. Therefore, the large scale model test of double-wall open-ended pipe pile needs further study urgently.

2　Large-Scale Model Tests

The maximum size of the existing model box is 2.0 m×2.7 m×2.4 m[12], and the minimum size is 50 cm×80 cm× 80 cm[13]. Most of the model boxes are composed of three sides of steel plate and one side of glass plate. The advantage is that the displacement of soil during the experiment can be observed from the side of the glass plate. The disadvantage is that there is boundary effect and size limitation. According to the boundary effect, the distance between pile and model box wall shall be greater than 7D(diameter of pile), and the distance between pile tip and model box bottom shall be greater than 4D. On this basis, the size of the model box designed is 3 m×3 m×2 m, with three sides of steel plate and middle and upper part of one side of glass plate, as shown in Figure 1.

Figure 1　Schematic diagram of model box

This project will develop the "track type CNC gear" cyclic loading device on the basis of the self-owned technology of the research group. The power device consists of reduction motor, DV300 converter, servo loading motor, hydraulic cylinder, high-pressure oil pump, pressure control box, static load test and analysis system and PLC control system. Three vertical motors with a power of 2.2 kW and a torque of 40 are selected for the reduction motor; One 5.5 kW and one 4 kW are selected for DV300 frequency converter, a 5.5 kW converter is installed on the main beam for adjusting two reduction motors on the main beam at the same time, a 4 kW frequency converter is used for adjusting the reduction motor on the secondary beam; gear and motor work smoothly together. The servo loading motor has a travel of 500 mm and a maximum output of 3.5 MPa, which can realize the cyclic loading mode of sine wave and M-wave;

The schematic diagram of structure of double-wall open-ended pipe pile is shown in Figure 2, and the physical diagram is shown in Figure 3. The pipe pile is composed of two concentric thin-walled circular pipes, both of which are 6063 aluminum alloy pipes, with Poisson ratio of 0.3 and elastic modulus of 72 GPa. The cross section dimension of the main part of the pile outer pipe is 140 mm × 134 mm (outside diameter × inside diameter). The cross section dimension of the main part of the inner pipe is 120 mm × 114 mm (outside diameter × inside diameter).

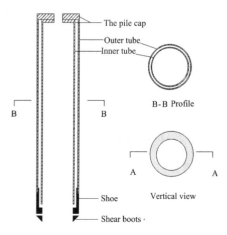

Figure 2 Schematic diagram of structure of double-wall pipe pile

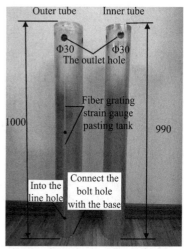

Figure 3 Physical diagram of inner and outer pipe

3 Conclusion

The dynamic response of open-ended pipe pile foundation under dynamic cyclic load of high-speed train, wind and wave is studied through large-scale model test. The specific technical indexes are as follows:

(1) Establish the mechanical model of pile lateral soil element considering lateral friction resistance, and construct the theoretical model of soil plug formation and cyclic weakening effect and its influence on horizontal bearing capacity of large-diameter pipe pile.

(2) Clarify the cyclic weakening of pipe piles under dynamic cyclic load, such as vehicle, wind and wave, and the macro-microscopic response of pile lateral soil.

(3) Clarify the macro and micro mechanism of the evolution of natural vibration frequency of single-pile support system, establish the theoretical analysis model of natural vibration frequency and amplification effect based on soil macro and micro variation.

(4) Construct a method for predicting the cyclic weakening and cumulative settlement of pipe pile under dynamic cyclic loading; develop visual software for engineering design and public safety disaster prevention.

References

[1] Paik K H, Lee S R. Behavior of soil plugs in open-ended model piles driven into sands[J]. Construction and building materials, 2005, 11(4): 353-373.

[2] Al-Douri, Riadh H., and Harry G. Poulos. "Predicted and observed cyclic performance of piles in calcareous sand."[J]. Journal of Geotechnical Engineering, 121. 1 (1995): 1-16.

[3] o'riordan n, allwright r, ross a,. Long term settlement of piles under repetitive loading from trains[C]// Symposium on Structures for High-Speed Railway Transportation, IABSE. Antwerp: [sn], 2003: 67-74.

[4] Huang Yu, Bai Jiong, Zhou Guoming. Model tests on settlement of a single pile in saturated sand under unilateral cyclic loading[J]. Chinese Journal of Geotechnical Engineering, 2009, 31(9): 1440-1444. (in Chinese)

[5] Zhu Bin, Yang Yongbiao, Yu Zhenggang. Field test of horizontal monotonic and cyclic loading of marine high pile foundation [J]. Journal of Geotechnical Engineering, 2012, 34 (6): 1028-1037.

[6] Chen Renpeng, Ren Yu, Chen Yunmin. Design method of pile foundation under vertical cyclic loading based on settlement control[J]. Journal of Geotechnical Engineering, 2015, 37 (4): 622-628.

[7] Huang Maosong, Liu Ying. Numerical analysis of vertical cyclic weakening of saturated clay pile foundation based on nonlinear kinematic hardening model[J]. Journal of Geotechnical Engineering, 2014, 36 (12): 2170-2178.

[8] Zhang Ga, Zhang Jianmin. Unified constitutive model and experimental verification of interface between coarse-grained soil and structure[J]. Journal of Geotechnical Engineering, 2005, 27 (10): 1175-1179.

[9] Bai Shunguo, Hou Yongfeng, Zhang Hongru. Analysis on critical cyclic stress ratio and permanent deformation of composite foundation improved by cement-soil piles under cyclic loading[J]. Chinese Journal of Geotechnical Engineering, 2006, 28 (1): 84-87.

[10] Fan Qinglai, Luan Maotian, Ni Hongge. Elasto-plastic effective stress analysis on soft soil foundation of large-diameter cylindrical structure subjected to cyclic loading[J]. Journal of Hydraulic Engineering, 2008, 39(7): 836-842.

[11] Yang Longcai, Guo Qinghai, Zhou Shunhua. Dynamic behaviors of pile foundation of high-speed railway bridge under long-term cyclic loading in soft soil[J]. Chinese Journal of Rock Mechanics and Engineering, 2005, 24(13): 2362-2368. (in Chinese)

[12] Ding Peimin, Xiao Zhibin, Shi Jianyong. Model tests of large driven pile jacked into sands and soil-pile interaction finite element analysis with compaction effects taken into account[J]. Industrial Construction, 2003, 03: 45-48.

[13] Zhou Jian, Deng Yibing, Ye Jianzhong. Experimental and numerical analysis of jacked piles during installation in sand[J]. Rock and Soil Mechanics, 2009, 04: 501-507.

(14) The Influence of the Number of Slots on the Damping Performance of the Stand-off Layer Damping Structure

Liang Longqiang, Huang Weibo*, Lyu Ping, Meng Fandi, Yu Chao

School of Civil Engineering, Qingdao University of Technology

Corresponding author E-mail addresses: huangweibo@qut.edu.cn

Abstract

Slotted stand-off layer damping treatments are presently being implemented in many commercial and defense designs. In order to explore the influence of the number of slots on the damping performance, the transfer function curves, modal frequencies and loss factors of the cantilever beams with slotted stand-off layer were tested and analyzed by hammer experiment. The results showed that when the number of slots increased from 2 to 10, the vibration response of the first three modes of the beams increased gradually, the loss factors became lower, which affected the damping performance. And the modal frequencies of the beans moved towards the low frequency as the stiffness of stand-off layer reduced.

Key Words

Slotted Stand-off Layer, Vibration Reduction, Damping Performance, Loss Factor

1 Introduction

Free layer and constrained layer damping treatments have been widely applied to reduce vibration in many fields ranging from architecture and ships to satellites[1,2]. In 1959, Whittier[3] proposed that the deformation of the damping layer could be further increased by adding a stand-off or spacer layer between the vibrating structure and the damping layer. In the ideal case, the stand-off layer simultaneously has infinite stiffness in shear and zero stiffness in bending[4-6]. This kind of damping treatment has proven to be effective, reliable and economical in a variety of environments.

To further reduce the structure quality and ensure the good damping performance, a new treatment with slotted stand-off layer was proposed and studied. Rogers and Parin[7,8] demonstrated experimentally that the treatment provided significant damping in aeronautical structures. Zhao[9] studied the noise reduction performance of high rail with slotted stand-off layer, and the results were very promising with reductions of the acoustic radiation pressure level of rail noises by 8 dB(A) by vertical and lateral excitation. Wei[10] found that the loss factor and damping performance of the stand-off layer damping structure were significantly improved, compared with constrained damping structures. On the basis, it is very important to study the influence of slot quantities on the damping performance of slotted stand-off layer damping treatment.

2 Experimental

2.1 Materials and Model Preparation

Stand-off free layer damping structure is composed of three parts: base layer, slotted stand-off layer and damping layer. In this experiment, Q235 steel plate was used as the base layer. The plate has density 7 800 kg/m³, elastic modulus 2.1×10^5 MPa and the size is 500 mm × 43 mm × 3.5 mm. Polyurethane foam was used as slotted stand-off layer, and the layer has density 170 kg/m³, size 500 mm × 43 mm × 12 mm. The damping layer adopted the damping rubber provided by Tianjin Rubber Industry Research Institute Co., LTD, with a size of 500 mm × 43 mm × 6 mm, a density of 1 500 kg/m³ and an elastic modulus of 45 MPa.

In order to improve the damping performance, the viscoelastic damping material (Qtech T501) developed by Qingdao University of Technology was used as the adhesive. The DMA results are shown in Figure 1.

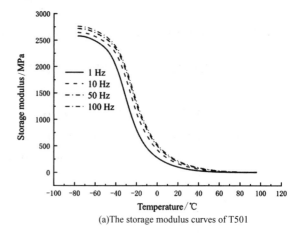

(a) The storage modulus curves of T501

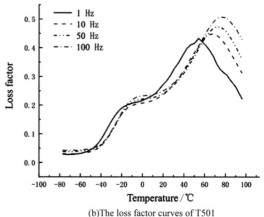

(b) The loss factor curves of T501

Figure 1　DMA figure of Qtech T501

The beams were prepared by hand scraping. The slots' depth and width is 6 mm and 30 mm respectively. The schematic diagram of the beam treated with slotted stand-off layer is shown in Figure 2, and the sample numbers of each group are shown in Table 1. The hammer test can be carried out after being cured at 25 ℃ for 72 hours.

Table 1　Model number and slot quantities of stand-off layer

Model No.	A_2	A_4	A_6	A_8	A_{10}
The slot quantities	2	4	6	8	10

Figure 2 Free layer damping structure with slotted stand-off layer

3 Results and Discussion

3.1 Amplitude-frequency Curves of the Transfer Function

The transfer function curves of stand-off Layer beams with different slot quantities were shown in Figure 3. The first three modes of cantilever transfer function are mainly concentrated in the range of 0～500 Hz.

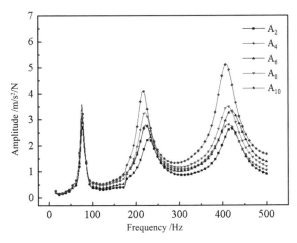

Figure 3 Amplitude-frequency curves of the transfer function

3.2 Test Equipment and Program

Figure 4 shows the hammer test system, including the exciting hammer, cantilever model, accelerometer, fixture, data acquisition instrument and analysis system. The exciting point and the pick-up point are 25 mm away from the fixed end and free end of the cantilever beam respectively, and the strike force is 50 N. The sampling frequency is 2048 Hz and the sampling number is 16384.

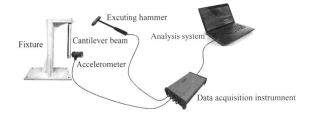

Figure 4 Test and analysis system

Vibration response peaks of the first three modes tend to get higher with the number of slot rising. The third-mode response peaks of A_2, A_4, A_6, A_8 and A_{10} are 2.68 m/s^2/N, 2.81 m/s^2/N, 3.34 m/s^2/N, 3.49 m/s^2/N and 5.13 m/s^2/N, respectively. The peak trend of the first and second mode is consistent with that of the third mode. This is probably because the more slots lead to lower bending stiffness of the stand-off layer, but it also significantly reduces the shear stiffness and weakens the effect of expanding the deformation of the damping layer. In addition, the contact area between the stand-off layer and the damping layer decreases, which reduces the vibration transmission efficiency from the base layer to the damping layer. Therefore, the vibration damping performance tends to be worse.

Furthermore, the third-mode response peaks of A_4, A_6, A_8 and A_{10} are 0.42 m/s^2/N, 0.66 m/s^2/N, 0.81 m/s^2/N and 2.45 m/s^2/N, respectively, higher than that of A_2. Compared with other models, the vibration response of A_{10} is more obvious, which indicates that the stand-off layer will gradually lose its function when slot reaches a certain quantity, and the vibration suppression performance of the structure will be seriously affected.

3.2 Frequency and Loss Factor

Table 2 shows the first three order modal frequencies and loss factors of $A_2 \sim A_{10}$.

Table 2   Modal frequencies and loss factors of $A_2 \sim A_{10}$

Model No.	First mode		Second mode		First mode	
	Frequency/Hz	Loss factor	Frequency/Hz	Loss factor	Frequency/Hz	Loss factor
A_2	79.22	0.1382	229.83	0.1537	418.29	0.1516
A_4	78.91	0.1119	228.60	0.1419	414.55	0.1402
A_6	77.63	0.1042	226.63	0.1398	412.31	0.1216
A_8	75.13	0.0929	224.47	0.1315	411.91	0.1023
A_{10}	74.28	0.0846	223.23	0.1230	407.62	0.1051

As the number of slot rises, the modal frequencies of the cantilever beams tend to be lower. These slots are somewhat equivalent to reducing the average thickness of the stand-off layer.

The bending modal frequency of a cantilever beam

$$f_n = \frac{\left(\frac{2n-1}{2}\pi\right)^2}{2\pi l^2}\sqrt{\frac{EI}{\rho S}} \quad (1)$$

When the thickness of stand-off layer changes, the parameter that determines the natural vibration frequency of the layer would be $\frac{I}{S}$. So

$$f_n \propto \sqrt{\frac{bh_v^3}{12S}} \quad (2)$$

That is

$$f_n \propto h_v \quad (3)$$

With the quantity of slot increasing, the average thickness of the stand-off layer decreases gradually and the modal frequencies of the beam get down. The loss factor of each order also decreases because of more slots, which indicates that the damping performance gets worse. That is consistent with the above analysis.

4 Conclusions

The number of slot has an important effect on the damping performance of cantilever beam treated with slotted stand-off layer damping. The more slots increase the amplitude of the modal response of the beam, reduce the loss factor, and the vibration suppression performance gets an obvious decline. In addition, the rising number reduces the bending stiffness of the layer, making the modal frequency getdown.

5 References

[1] Wang Guoqing, etal. "The Application of Viscoelastic Damping Material on Vibration and Noise Reduction." Science Technology and Engineering 3(2013):89-96.

[2] Huang Weibo, Zhan Fengchang. "Studies on the dynamic mechanical and vibration damping properties of polyether urethane and epoxy composites." Journal of Applied Polymer ence 50. 2(1993).

[3] Whittier, J. S., "The effect of configurational additions using viscoelastic interfaces on the damping of a cantilever beam." Wright Air Development Center 58-568 (1959).

[4] Yellin, Jessica M. "An analytical and experimental analysis for a one-dimensional passive stand-off layer damping treatment." Cancer Research 66. 4(2004):295-306.

[5] Jessica M. Yellin, I. Y. Shen, and Per G. Reinhall. "Analytical model for a one-dimensional slotted stand-off layer damping treatment." Proceedings of Spie the International Society for Optical Engineering 3989 (2000):132-141.

[6] Jessica M. Yellin, and Iyen Shen. "Analytical model for a passive stand-off layer damping treatment applied to an Euler-Bernoulli beam." Proceedings of Spie the International Society for Optical Engineering 3327 (1998).

[7] Parin M, Rogers L C, Falugi M, et al. "Practical stand-off damping treatment for sheet metal." Proceedings of Damping' 89. West Palm Beach, Florida: Wright Laboratory Flight Dynamics Directorate, (1989): 1-26.

[8] Rogers L C, Parin M. "Experimental results for stand-off passive vibration damping treatment." Smart Structures & Materials' 95. Mishawaka, Indiana: International Society for Optics and Photonics, (1995): 374-383.

[9] Zhao Caiyou, Wang Ping. "Theoretical analysis and experimental investigation on cross-legged silent rail." Journal of Southwest Jiaotong University, 2 (2013): 290-296.

[10] Wei Zhaoyu, Shi Xiuhua, Li Shichao, et al. "Modeling and simulation of a constrained beam with slotted stand-off layer." Mechanical Science and Technology for Aerospace Engineering, 7(2009) : 876-880.

3. Environmental Engineering and Comprehensive Utilization of Resources

(1)Rewetting of Fens: a Case Study from Northern Germany

Florian Schlomo Hetzel, Petra Schneider

Department of Water Management, University of Applied Science Magdeburg-Stendal

Abstract

Peatlands have been drained for several decades to make them usable for agriculture or forestry. The biotopes, originally rich in species, have been changed in an irreversible way causing negative consequences, particularly to their soil structure, an altered water balance and large CO_2 emissions. The peat bodies formed over several hundred years were mineralized by dewatering or directly mined peat material in a short time. Many of these areas continue to be used for agricultural purposes although the soil structure continues to degrade. Therefore, an alternative use is necessary. It is advisable to cultivate plants on these sites which both preserve the natural structure of the moorland and produce profitable yields (for example paludiculture). For the implementation of such projects, the drained areas must be rewetted. Since each area has to be considered individually, water management is a core task of such projects. The following research considers a concept for the rewetting of an 8.5 ha large area in Mecklenburg-Western Pomerania (Germany) based on the bachelor thesis Development of Rewetting Possibilities of a Degraded Moor in an Area of the Polder Bargischow by F. Hetzel.

Key Words

Fen, Peatlands, Soil Structure

1 Introduction

A peatland is an area in which peat is produced by a long-term water-saturated soil. Anaerobic conditions in the subsoil and peat-producing corps are responsible for the peat accumulation. According to hydrological considerations, there are two main types of peatlands: A bog is supplied only by rainwater (ombrogenous), so it has a lower concentration of nutrients then a fen that is dominated by surface and ground waters. Thus it is possible to cultivate a fen with more different plants[1].

Originally, 4.5% of the land area of Germany was covered with peatlands which were drained by 95%[2]. In the north eastern Federal State Mecklenburg-Western Pomerania the proportion of peatlands is even 12.7% and about 99% of these wetlands are fens. The melioration in Germany was a continuous process through which the areas have been cultivated as agricultural land. Since the soil had to be drained for the cultivation of crops, the natural function of the peatlands has been destroyed.

Three important reasons underline the need for rewetting of drained peatlands: climate change mitigation, costs and new possibilities in the field of agricultural use. Today it is known that the natural functions of a peat area have positive effects on the climate as greenhouses gases are kept in wet peatlands. Among other things, carbon, phosphorus, nitrogen and sulfur are stored in a peatland. Thus, a peatland is a nutrient sink. In addition, wetlands act like sponges and prevent flooding[3]. In some areas of Germany, the maintenance costs for drainage of the peatlands are now higher than the yield to be generated. The agricultural use of wetlands is no longer profitable in view of conventional cultivation.

Drainage in peaty soils leads to sacking, shrinkage and swelling, humification, peat shrinkage and shifting and leaching[4]. These processes are often irreversible, which is why the use of the peatland without drainage should be sought. Plants such as reed, black alder or cattail (Typha) are alternative uses. For the economic use of wet and rewetted peatlands in the sense of forestry or agriculture, the term paludiculture is therefore developed by the University of Greifswald (Germany). The aim is to use the biomass of plants that form peat or maintain peat. According to current knowledge, for example cattail is a peat-preserving plant that can be used as a building material, for energetic purposes or for food production. Further types of uses are under investigation. However, the cattail and other plants growing on fens are not yet recognised as an agricultural crop in Germany and are therefore not eligible in terms of financial support. Furthermore, the biotope protection regulation prevents a targeted attachment of cattail.

A rewetting of fens has only been carried out for a few years, which is why there is still room for innovations. The conception of water management is therefore a major task in rewetting, where each area must be considered as an individual case study. There is no general solution and the adjacent area is often affected by the rewetted area, so the natural conditions can often no longer be restored. Furthermore, surface waters are usually no longer in a natural state and the water balance is altered.

2 Case Study in Northern Germany

2.1 Location

A rewetting concept was planned as a pilot project targeting paludiculture for an approximately 8.5 ha large degraded peatland in the north east of Germany. The area is enclosed by earth dams to the west, north and east. In the south it borders another drained peatland area. The northern dike is dammed up, forcing the water to seep through the dike at a certain water level. The water levels depend on the level of the nearby Peene river. The scope of the rewetting project is to use the area for cattail cultivation, which is already growing there naturally. The average water level should be 5 to 15 cm above the ground. Since the existing machines can only mow when the area is free of water, it must be possible to drain the area. Further, the adjacent neighbouring property should not be affected by the rewetting measures.

2.2 Methodology and Data Base

The investigation strategy comprised a site investigation with collection of historical data, the calculation of a water balance, and the preparation of a feasibility analysis of potential rewetting strategies. Germany-wide archived data can be used to provide initial information about the area, for example soil overview maps, moorland site catalogues and drilling samples from the state drill data archive. Furthermore, digital elevation models are available in Germany, which can be purchased from the competent authorities.

2.3 Hydrological Boundary Conditions

An important parameter for rewetting is the soil structure. During the soil development of fens, the drainage causes a peat shrinkage. For the area under investigation this leads to a shrinkage below the adjacent surface water (Peene river)[5,6]. Figure 1 shows the course of a fen during dewatering due to its degradation state.

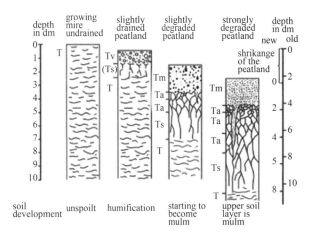

Figure 1　Shrinkage of peatlands through dewatering (Source: changed image of Paludikultur 2016[7])

In Germany, usually a fen shrinkage of 5 to 10 mm per year is used for further calculations in case of grassland. Figure 2 shows a height model of the project area and surroundings. It is clearly visible that the area is below the water level of the adjacent watercourse (dark green and light green sections). It is also evident that the relief is heterogeneous.

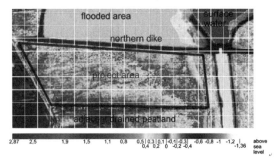

Figure 2　Height model of the project area (source: modified digital elevation model 2 by F. Hetzel)

The water balance is determined by the factors for water increase (groundwater flow, precipitation, etc.) and water withdrawal (evaporation, runoff, etc.). The groundwater flow can be determined at existing groundwater measuring points. As a rule, the infrastructure of the groundwater monitoring network is well positioned in Germany, but the distances between the monitoring locations points are too large in the project area. Therefore, it is necessary to install water level meters for more accurate data. In addition, a special feature of the project area is the probability that seepage water leaks on the area. This is due to the higher water level of the adjacent water body and a dike in the north of the project area that is not built for this case. Due to a broad meteorological monitoring network, the precipitation data can be taken from the surrounding measuring stations. The data for evaporation can also be taken from the meteorological measuring stations. In order to calculate the real evaporation of the future dammed up surface water level, the Dalton method is suitable. For the calculation of the discharge through existing drainage trenches, the cross-sections of the trenches as well as the data of the water levels are necessary. Finally, a comparison of the water inflow and outflow gives an indication if the area is supplied sufficiently with water all year round. For larger areas, the use of a simulation program is advantageous[8].

3 Results

3.1 General Technical Approaches

There is no general technical solution for rewetting. In the case of the project area, the following approaches are taken into account:

• profiling the soil surface (to obtain a uniform area),

• soil loosening (if the soil is compacted by degradation),

• backfilling or decoupling the drainage ditches, and

• construction of a dam to maintain a water level.

In addition, a device is required to drain the area.

A cost calculation is an important feasibility criterion for selecting the type of rewetting measures. The costs should be reasonable, because if the project is successful, an incentive for farmers from degraded fens can be created. Furthermore, the decisive factor in Germany is which subsidies are possible and to what extent.

3.2 Results for THE Project Area

For cost reasons, it is not feasible to conduct a profiling of the area. Drainage could be prevented by a wooden sheet pile wall through the drainage trench, while the trench could also be used for drainage during the crop harvesting period. A soil wall north of the drainage ditch could be erected as a dust-free structure with a drain to be installed at the deepest point for draining. In order to collect information on the site specific impacts of the proposed rewetting methods, the experimental soil loosening using high-pressure lances could be applied. No research is done on that subject yet.

Since such a project has to go through many steps of approval procedures, like Environmental Impact Assessment (EIA), the implementation is often delayed due to a lack of experience and knowledge. For this reason, a cooperation of different specialists is necessary. Due to legal uncertainties, no practical measures have been implemented yet in the pilot area.

4 Conclusions

At the case study site, there is no final rewetting solution yet, due to a high number of influencing conditions. For an exact determination, more detailed investigations are needed, as for instance drilling samples must be taken on site.

Even if each area has to be considered individually, the success of pilot projects is important. Ecological and economic aspects must be taken into account to achieve rewetting by means that are easy to implement. In case of a rewetting success, peat-friendly agricultural management becomes feasible. Furthermore, there is the possibility to promote the regional area in Germany for a development of value chains, as for example cattail flasks to pellets, insulation material or food or reed for thatched roofs on site. A smart communication with the surrounding population is essential for a successful implementation, as rewetting can only take place in cooperation with the people affected.

References

[1] Hetzel F., 2018. Development of rewetting possibilities of a degraded moor in an area of the polder Bargischow, bachelor thesis, University of Applied Science Magdeburg-Stendal, Germany.

[2] Moore in Deutschland, available: https://www.nabu.de/natur-und-landschaft/moore/deutschland/index.html. Accessed Aug 24, 2018.

[3] Klimaschutz, available: https://www.bundesregierung.de/Content/DE/Artikel/2014/08/2014-08-14-sommermoore.html. Accessed Aug 23, 2018.

[4] Schmidt W., 1981. Kennzeichnung und Beurteilung der Bodenentwicklung auf Niedermoor unter besonderer Berücksichtigung der Degradierung, F/E-Bericht, Institut fuer Futterproduktion Paulinenaue, Germany.

[5] Succow M. and Joosten H., 2001. Landschaftsökologische Moorkunde, E. Schweitzerbart'sche Verlagsbuch, Stuttgart, Germany.

[6] What is peat? available: http://www.peatsociety.org/peatlands-and-peat/what-peat. Accessed Aug 22, 2018.

[7] Wichtmann W., Schroeder C., Joosten H., 2016. Paludikultur - Bewirtschaftung nasser Moore, Schweizerbartsche Verlagsbuchhandlung, Stuttgart, Germany.

[8] Wieder R. K. and Vitt D. H., 2006. Boreal Peatland Ecosystems, Springer-Verlag, Berlin Heidelberg, Germany.

(2) Formation and Succession of Oxbow Lakes in the Middle Elbe Biosphere Reserve

D. J. Hunger

Department of Water and Waste Water Management University of Applied Science Magdeburg-Stendal

Abstract

Oxbow lakes are part of the natural development of rivers. Usually they develop during flood events and further run through three different stages of succession—the initial, optimal and terminal stage. In each of these stages the characteristics change in terms of e.g. bed material, water depth and coverage by aquatic plants. This influences the colonization by e.g. macroinvertebrates and fishes, resulting in species that are adapted to the characteristics of the different stages.

Normally, the succession of oxbow lakes is a process over centuries. Anthropogenic impacts accelerate the succession, e.g. by nutrients and organic pollution, and hamper the development of new oxbow lakes, e.g. by straightening the main channel of the river. These circumstances pose a threat to endangered species because oxbow lakes are a very important and rare habitat. Hence, it's reasonable to pay more attention to oxbow lakes and restore them by desludging or reconnecting.

Usually, it's difficult to classify this waterbody as a lake or river because some of them have a floating velocity and show typical characteristics of a river, whereas some of them seem to be lakes, with no flowing water and the typical hydrophytes of lakes. These phenomena of waterbodies are under pressure of a natural deterioration. The next important step is to develop a guidance to evaluate the status of oxbow lakes as a fundamental step before starting revitalization.

Key Words

Oxbow lakes, Flood, Succession, Waterbody

1 Formation of Oxbow Lakes

The natural formation of oxbow lakes relates to the dynamic of the flowing river. Oxbow lakes appear in the lower course of rivers where the flow velocity and the slope are low[1]. The water in the river channel flows faster around the outside bend and erodes the bend of a river, whereas the slowly flowing water on the inside bend deposits the sand and mud. These conditions lead the river to form curves through the landscapes (meander) With the time the meanders change in shape and grow.

At one point the curves of the river become very close and the river breaks through this tin barrier (Figure 1). The river no longer flows through the meander but straight along the new channel. The old course of the river becomes an oxbow lake and will naturally silt during centuries like a normal lake. They are crescent-shaped (like a bow) and right after the formation as wide as the main river. Besides cutted meanders, oxbow lakes also form after a flood, when the river changes the river course and the water remains in tidal pools.

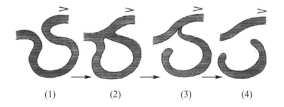

Figure 1 (1) Meander (2) oxbow lake, both sides linked to the main channel (3) one side linked to the main channel (4) no connection to the main channel [2]

A non-natural way of how oxbow lakes form is by man-made river training. It's often used to navigable waters by straightening the water body or building levees to increase the flood control. They are leftovers of a time where the river wasn't forced in the constructed channel and had the opportunity to use its dynamic to change the landscape naturally.

A very important characteristic of an oxbow lake is the connection to the main channel. The connection has an influence on the water quality, solved oxygen, nutrient balance and species, especially the appearance of macroinvertebrates, fishes and hydrophytes.

2 Development Stages of Oxbow Lakes

After the oxbow lake is formed, it is running through three stages. Each stage has its own characteristics and is important for a special kind of biota. The main influence of the stage has the connection to the main channel and whether there is a levee which aggravates the reconnection and the water exchange during flood. When the flow velocity decreases down to zero, the oxbow lake starts the transformation to standing water body. Especially in the summer time the temperature of the standing water body is much higher in comparison to the temperature of the river. The biomass production increases, as well as the phytoplankton and zooplankton population, because they will not be swept away by flood.

The high amount of biomass and nutrition from the river, which are temporary washed in, cause natural eutrophication. Plants take their chance to grow from the banks into the oxbow lake and the "biogenic silting" takes place.

How fast the successional process proceeds depends on several aspects:

- size and depth of the water body
- sedimentation
- nutrition
- water level fluctuation

Initial Stage

Figure 2 Oxbow lake in the initial stage with connection to the main channel but sedimentation begins to cut off the connection[2]

The initial stage is characterised by mesotrophic nutrition value with no sludge on the ground and a large surplus water. Often there is still a connection to the main channel and the oxbow lake and its vegetation are like the river, but the flow velocity is already steady. At the river bank cane brake and first pioneer vegetation grow.

Optimal Stage

Without a proper connection to the main channel, the optimal stage is characterised by a sludge layer on the ground with high nutrition value. If the backwa-

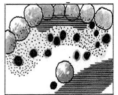

Figure 3 Oxbow lake in the optimal stage with floating leaves on the water body and cane brake growing on the bank[2]

ter is behind a levee, also a flood has no effects on the water body. These conditions affect the water depth and lead to high species-rich vegetation and wildlife. Bacterial activity within the sludge consumes a lot of the oxygen and during the summertime a lack of oxygen can be noticed.

Terminal Stage

Figure 4 Backwater in the terminal stage with advanced siltation and numerous species of cane brake

The terminal stage is characterised by polytrophic nutrition value and a small, shallow and rapidly warming water body, which is almost completely covered with floating hydrophytes. The siltation process is on its maxima and the vast sludge layer supports plants to grow "into" the backwater. Small bosk shows up at the former river bank and the floodplain forest develops.[3,4] In this stage the oxbow lake is basically a moor.

3 Habitat for Numerous Species

If backwaters aren't under anthropogenic pressure the wildlife in such kind of habitat can be rich. Each stage has its own compound and is connected to the characteristics of every stage. The bank of a backwater is mostly covered with cane brake like spike rush (Eleocharis palustris). The water body is covered with floating hydrophytes, such as yellow water lily (Nuphar lutea) and white-water lily (Nymphaea alba). Several dragonfly species, water beetles, macroinvertebrates and mussels are using the backwaters for hiding, reproduction and food source. This diversity of species also attracts superior animals, e. g. birds like the kingfisher (Alcedo atthis) or the bearded reedling (Panunrus biarmicus).

Some fishes also use the backwater as a refugium to spawn, to feed and to escape from the energy sapping life in the open river. Those like the bitterling (Rhodeus amarus), crucian carp (Carassius vulgaris) and bream (Abramis brama) prefer the condition they find in the optimal stage of a backwater with standing water and lots of zooplankton[4,5].

4 Revitalization Methods

To plan there vitalization it is necessary to know the condition of the oxbow lake. Although a guidance to evaluate the status of rivers or lakes exists in Germany, there is nothing like this for oxbow lakes.

The main goal in revitalization of oxbow lakes is to affect the water body in terms of nutrition, oxygen consumption in the sludge, and flow velocity. As shown in Chapter 2, these qualities are the major catalysts for the successional process.

There are two common methods of oxbow lake-revitalization. One possible method is to restore the connection by soil excavation and construction of rock ramps between the backwater and the main channel[6]. The water exchange with the main channel is recovered and the flow velocity will rise. That tends to result in minor sludge layer and less oxygen consumption by bacteria. Also, the nutrition value will decrease. To accomplish these effects desludging the water body is another possibility. In combination with diversifying hydro ecomorphological structures, species number of macroinvertebrates and plants may increase steadily[7,8]. Nevertheless, depending on scale of a project and amount of sludge, both methods are very expensive.

5 Conclusions

A river needs enough space to spread

out and to form oxbow lakes. The artificial stabilization of the river banks and the man-made river training inhibit the formation of oxbow lakes. But they are important habitats for endangered species and can have positive side-effects on the main river channel. Revitalization is of avail for water body conservation, but it is expensive.

Nevertheless, natural siltation will continue and one day oxbow lakes might be wiped off the map. The next important step to take must be the development of a uniform guidance to evaluate the status of oxbow lakes. It is fundamental to know the conditions, otherwise the revitalization methods might not be productive.

References

[1] BÖTTGE, G., 2009. Entstehung und Sukzession von Altgewässern. Flussaltwässer. Ökologie und Sanierung. Wiesbaden: Vieweg + Teubner Verlag / GWV Fachverlage GmbH Wiesbaden.

[2] DWA-M 607, 2010. Altgewässer - Ökologie, Sanierung und Neuanlage. Hennef (Sieg): DWA. DWA-Regelwerk. M 607.

[3] Lüderitz, V. Kunz, C. Langheinrich, U. 2009 Flussaltwässer, Ökologie und Sanierung Vieweg + Teubner Verlag/GWV Fachverlage GmbH, Wiesbaden.

[4] Remy, D., 2009b. Lebensraum Altwasser im Initila-, Optimal- und Terminalstadium. Pfanzen. Flussaltwässer. Ökologie und Sanierung. Wiesbaden: Vieweg + Teubner Verlag / GWV Fachverlage GmbH Wiesbaden.

[5] Zupke, U. Reichhoff, L. 2015 Fische und ihre Habitate in Auenstillgewässern an der Mittelelbe. Naturschutz im Land Sachsen-Anhalt (52), p. 45-62.

[6] Seidel, M. Gewässermorphologische Integrität von Flüssen. Wiederanschluss von Altgewässern zur Verbesserung der Lebensbedingungen flusstypischer Arten. Korrespondenz Wasserwirtschaft, p. 107-111.

[7] Seidel, M. Voigt, M. Langheinrich, U. Hoge Becker, A. Gersberg, R. M Arevelo, J. R. Lüderitz, V. 2016 Reconnection of oxbow lakes as an effective measure of river restoration WILEY-VCH Verlag GmbH & Co. KGaA, Weinheim.

[8] Langheinrich, U. Dorrow, S. Lüderitz, V. 2002 Strategies for protection and conservation of surface waters in floodplain landscapes of the Middle Elbe Biosphere Reserve. -Hercynia 35: 17-35.

(3) The Elution of Heavy Metals at Tailings Piles

Sven-Simon Lattek

Department of Water Engineering, University of Applied Sciences Magdeburg-Stendal

Abstract

Tailings piles are indispensable waste disposal facilities resulting from mining in the potash industry that are constructed as complex engineering buildings. Pile waste water can influence the pH value, which can lead to an acidification of the ground. To analyse this process, a column experiment with different sample soils has been used. This experiment has confirmed that the elution of heavy metals can be influenced through the pH value.

Key Words

Heavy Metals, Tailings Piles, Elution

1 Introduction

The physico-chemically behaviour of the pile body is affected by the remains itself, the underground conditions as well as the climatic conditions of the particular company location[1].

Pile waste water develops from rainfall infiltration as well as gravitative free and remaining humidity, which percolates in the underground and affects the pH value or rather lowers the pH value. Through this acidification of the ground, the heavy metals and trace substances that are adherent to the soil matrix, are eluted and consequently bioavailable[2,3].

2 Analysis

During this research, the elution of heavy metals and trace substances have been analysed. Therefore, soil samples from K+S locations in Zielitz (Saxony-Anhalt, Germany) and Hattorf (Lower-Saxony, Germany) have been gathered. Furthermore, the analysis addressed the question to what extent the acidification of the ground can be prevented by adding limestone[4].

Therefore, an experiment with columns has been built. The Perspex columns have a drain system and contain sample soils in the initial state as well as mixtures of soil and limestone. The columns have been watered with pile waste water. The gained eluate has been regularly analysed regarding the pH value and the heavy metals, trace substances and salinity. The limestone has been either admixed in the samples homogeneously or as a lime column.

3 Conclusion

The experiment confirmed that through the percolation of the pile waste water heavy metals eluate and become bioavailable. It is possible to stabilize the pH value and consequently to reduce the elution of heavy metals significantly by adding limestone. Furthermore, it has been discovered that the elution of heavy metals lowers, and the pH value rises over time, which implies that the heavy metal output is finite.

Additionally, it has been determined that the pH value decreases and therefore consequently the elution of heavy metals rises during the transition from a soil sample containing limestone to a reactive soil sample (without lime addition). Therefore, it can be concluded that the overall area of the reactive soils must be combined with limestone to stabilize the pH value and to minimize the elution of heavy metals.

References

[1] Alloway, B. J., ed., 2013. Schwermetalle in BÖden: Analytik, Konzentration, Wechselwirkungen. Berlin: Springer.

[2] Hirner, A., Rehage, H.-O., and Sulkowski, M., 2000. Umweltgeochemie: Herkunft, Mobilität und Analyse von Schadstoffen in der Pedosphäre. Darmstadt: Steinkopff.

[3] Jasmund, K. and Lagaly, G., 1993. Tonminerale und Tone: Struktur, Eigenschaften, Anwendungen und Einsatz in Industrie und Umwelt. Heidelberg: Steinkopff.

[4] Rauche, H., 2015. Die Kaliindustrie im 21. Jahrhundert: Stand der Technik bei der Rohstoffgewinnung und der Rohstoffaufbereitung sowie bei der Entsorgung der dabei anfallenden Rückstände. Berlin, Heidelberg: Springer.

(4)Electronic Waste—Challenges and Chances

Sven Mauch

Recycling and Waste Management, University of Applied Sciences, Magdeburg - Stendal

Abstract

Electronic devices such as smartphones, computers, gadgets and etc. are commonplace in every household worldwide.

Due to the development and progress in technologies the lifespan of such devices are shortening.

At the same time the flood of electronic waste expands and humanity faces new challenges.

How can an electronic waste infrastructure be implemented?

How is it possible to dispose electronic waste and protect the environment?

How can the rare and expensive natural resources used in e-devices be recovered and reused?

How can illegal disposing and recycling structures be stopped?

The list of unanswered questions is long. But besides these problems the topic also provides big chances in protecting natural resources. It opens doors for steps in the direction of an efficient and social way to handle this kind of waste.

Companies start changing their policies on how to handle old devices and introduce innovative recycling systems. At the same time countries start to implement electronic waste infrastructures.

Despite the developing efforts the way to a well-functioning system is still a long one to go.

The text describes the challenges to be faced and progresses that have already been made in this field.

Key Words

Electronic Waste, Electronic Devices, Lifespan

1 Introduction

Electronic waste includes all discarded electronic devices even the ones that are meant to be recycled or reused.

The worldwide demand for electronic devices increases constantly. At the same time the price and the lifespan decrease.

Because of this condition the amount of produced electronic waste is high all time.

Such rapid development led to an

overtax of the waste managing economy. A big amount of the E-Waste is illegally disposed and damaging the environment and human health.

Only small parts get officially collected. The amount of recycled E-Waste is even lower and tons of rare raw materials, such as gold are not reused.

2 Problems of Electronic Waste

(1) Industrial countries: Industrial countries have two main problems. First the consumption of E-Devices increased rapidly during the last decade. Due to a good economic situation most of the people own more than one electronic device. For example a computer, a smartphone and a tablet. Also short product replacement cycles make users discard the devices long before they actually break. Another problem is an increasing variety of new electronic devices and innovations. One example is cloud computing which creates big data centers. The second problem in industrial countries is the bad collecting and recycling quote. Compared to other sorts of waste the recycling of E-Waste is far to less developed. An appropriate infrastructure is missing and a big part gets illegal disposed in most industrial countries.

(2) Problems in developing countries: Developing countries also fight with two big problems. The first and bigger problem is the disposal of worldwide E-Waste in these countries. Due to a bad legislation system for E-Waste developing countries suffer under a big import of foreign waste.

The infrastructure is bad and most of the garbage gets disposed under tremendous bad environmental conditions. Another part gets recycled by the informal sector mostly without any health protecting possibilities. Another problem in developing countries is the increasing consumption of electronic devices. Lower prices are making it easy to afford devices like smartphones and computers.

3 Technical Recycling Methods

(1) Separation and recycling of the raw materials: E-Waste contains gold, silver, copper, platinum, palladium, iron aluminum and many more valuable and technically recoverable materials. This high variety of materials makes it hard to separate and recycle every single one of them. Modern facilities use different methods and only concentrate on single groups of materials. That makes it hard to picture every single recycling method. A big part of the E-Waste recycling industry concentrates on the reuse of copper. Compared with most of the other materials in electronic devices, copper is easy to separate and can be recycled without losing any quality. In general this process can be split into the following work steps:

The waste gets crushed in a shredder. Afterwards iron parts get sorted out. This can happen through magnets. In the next steps plastic and aluminum pieces get segregated. The leftover materials are gold, silver and mainly copper[1,2].

These materials get crushed in a shredder again. Afterwards they get melted. The result of this step is copper with a comparable bad purity. To solve this problem the copper follows an electrolysis. Through this process materials like gold and silver get segregated and mixed in a melt. The cooper remains in nearly 100% purity. The mixed melt of gold and other materials gets further processed to filter the expensive raw materials.

One problem of this recycling method is that all materials get crushed together at the beginning of the process. Separating the E-Waste into modules with similar characteristics would help to improve the efficiency of the recycling process.

(2) Progressing techniques: Due to the fact that the separation of E-Waste turned out to be very complicated, companies and scientists started to develop helping techniques. For one of these technologies the power of high electronic shocks is used to split phones into several parts. The diversity of different materials in these segregated parts is far smaller than before what helps to simplify the recycling process. Another problem of the common E-Waste recycling process is the separation of gold, silver and other rare metals. Many companies use hydrochloric acid to obtain the materials. But the process is environmental unfriendly and not too efficient. One method to solve this problem is a newly identified chemical connection. It develops strong complexes of the different metals. These complexes are easy to separate. At the same time the method is environment-friendly[3-5].

4 Conclusion

Electronic waste is a problem that makes the world face a lot of challenges. The way to a functional recycling system is still a very long one and the biggest part of the garbage ends up in countries without any recycling facilities. But if the implementation of a well working E-Waste recycling system succeeds, the positive sides of the topic could overweight the negative ones. Rare natural resources could be reused and the environment could be protected. Nevertheless the biggest problem is in the consumption of electronic devices. The "throw away society" mentality has to change. Otherwise the recycling industry won't have a chance to go after the huge amount of E-Waste.

References

[1] Balde, C. P., Forti, V., Gray, V., Kuehr, R., Stegmann, P., 2017. The Global E-Waste Monitor 2017.

[2] Nicolai Kwasniewski, 2017. Planet Elektroschrott, Spiegel Online.

[3] Matze Brandt, 2015. Beim großten Elektroschrott Recycler der Welt, MDR. DE Einfach genial.

[4] Faszination Wissen, 2016. Warum Elektroschrott ein Problem ist, Bayerischer Rundfunk.

[5] Heidi Wilhof, 2018. Uberblick von Agbogbloshie.

(5) Urban Mining—Discovering the Values of Construction Products

Julia Marie Zigann

University of Applied Sciences, Magdeburg-Stendal

Abstract

The treatment measure of building materials has not yet reached its full potential. An Ecological Evaluation has made it possible to construct a comparison between reuse and recycling or disposal. This evaluation is carried out exemplarily on brick waste. A better end result is achieved with regard to the greenhouse effect for the reuse of brick than the recycling/disposal. This academic presentation for my current research describes the waste hierarchy of Germany, waste generation and the Ecological Evaluation.

Key Words

Urban Mining, Construction Products

1 Introduction

Our cities contain several resources in buildings and infrastructure, including but not limited to: mineral building materials (cement and bricks for example), various metals and structural components (stairs and dividing walls in offices).

While dissimilating these man-made products the construction industry produces the largest percentage of waste in Germany. In 2014 they managed to recover 89.5 percent[1] thereby becoming a main provider of raw materials.

The treatment measure of building materials has not yet reached its full potential. Due to combining material with impurities brought about by construction and waste, they are often used in low profile projects, road construction for example.

Steady high demands for raw resources in construction are predicted. These raw materials, like gravel and sand, which are used for general construction as well as civil engineering, are unattainable in environmentally sensitive areas, which should be protected and preserved. Worldwide resources are being stripped at an alarming rate, adding to losses in biological diversity within nature reserves, moorlands, bodies of water, rainforests, and so on.

2 Building Material Deconstruction Projects Compared to Recovery and Disposal

Lasting change in Germany should be implemented within the waste disposal hierarchy of waste management ("Kreislaufwirtschaftsgesetz" KrWG). The goal is to fulfil cost benefit (§ 7 KrWG) needs while prioritizing the environmental care and conservation. That includes targeting the high value recyclable raw materials in construction, not just material that can be used in low profile projects, like the aforementioned road work. There are exceptions however, prioritizing low profile material due to economic constraints and needs.

There are pilot projects that successfully implemented usage of recycled higher value materials, creating wins for participating parties. The pre-existing requirement is that it is planned as a raw material orientated demolition and recovery venture. These projects can be carried out with the help of experienced construction office personnel.

3 Ecological Evaluation

The software "Umberto" enables life cycle analyses. In a workplace a system for ecological evaluations is implemented, which enables a comparison between reuse and recycling or disposal. The individual material and energy flows are evaluated based on usage categories, in this case the greenhouse effect. The functional unit of measurement used to describe ecological evaluation results is 1 m^3 bricks. The indicator used to characterize the results is GWP (Global Warming Potential) in kg CO_2-equivalent.

4 Conclusion

Life cycle analyses are the only internationally normalized method to approximate lasting effects a product will have on the environment[2]. Future growth in this area is expected in the economy, political scene and general society for the sake of "Development and improvement of products, strategic planning, political decision making, marketing, and so on"[2,3].

Responsible management of resources cannot completely put a stop to global warming, but it can minimize potentially irreversible damage[4].

Increasing population size coupled with increased difficulty in accessing raw materials make finding solutions in resource management essential. Safeguarding our natural resource access is made all the more crucial in the face of modern technology[5].

Old buildings and infrastructure are raw material storage spaces that come with immense potential. High profile waste management is not yet implemented as it could be[6]. High potential for recycling valuable materials lies in the "construction site waste areas", 96.9 percent of which is recovered and used. Some people still maintain that reusing building site material is not economically sound. Nevertheless, in our field especially, we have a social obligation to fulfil on behalf of future generations and the environment.

References

[1] Bundesverband Baustoffe - Steine und Erden e. V., 2017. Mineralische Bauabfälle Monitoring 2014. Berlin. Druckwerkstatt Lunow. Online verfügbar unter http://kreislauf-wirtschaft-bau. de/Arge/Bericht-10. pdf. Open website: 10. 04. 2018. pp. 11, 19.

[2] Klöpffer, Walter, Grahl, Birgit, 2009. ökobilanz (LCA). Ein Leitfaden für Ausbildung und Beruf. Weinheim: Wiley-CH Verlag GmbH & Co. KGaA. pp. 2 zit. nach ISO 14040; 12.

[3] BMUB, 2016. Deutsches Ressourceneffizienz-programm II - Programm zur nachhaltigen Nutzung und zum Schutz der natürlichen Ressourcen. 1. Aufl. Berlin: Druck- und Verlagshaus Zarbock GmbH & Co. KG. Online: http://www. bmub. bund. de/fileadmin/Daten_ BMU/ Pools/ Broschueren/progress_ii_broschuere_bf. pdf. Open website: 28. 02. 2018. pp. 66.

[4] IPCC, 2014. Klimaänderung 2014. Synthesebericht. Online verfügbar unter https://www. ipcc. ch/ pdf/ reports-nonUN-translations/deutch/ IPCC-AR5_ SYR_ barrierefrei. pdf. Open website: 15. 04. 2018. pp. 19.

[5] BMUB, 2015. Deutsches Ressourceneffizienzprogramm (ProgRess). Programm zur nachhaltigen Nutzung und zum Schutz der natürlichen Ressourcen, 29. 02. 2012. 2. Aufl. Paderborn: Bonifatius GmbH. https:// www. bmu. de/fileadmin/Daten_ BMU/Pools/Broschueren/ progress_broschuere_ de_ bf. pdf. Open website: 10. 04. 2018. pp. 20.

[6] Dechantsreiter, Ute, 2018. Calling at 23. 03. 2018. Abbruchunternehmen. com, o. J. Abriss, Abbruch und Rückbau im Detail. Online verfügbar unter http://www. abrissunternehmen. com/abriss/. Openwebsite: 26. 04. 2018.

(6)The Elimination of Phosphorus on Sewage Treatment Plants

Weber Vincent

Department of Water, Environment, Construction and Safety,

University of Applied Sciences, Magdeburg-Stendal

Abstract

The eutrophication of waters was discovered after the beginning of wastewater treatment and in the eighties became the focus of sanitation. In the Cleaning Agents and Detergents Act, maximum levels have been set for phosphate in order to minimize the input of phosphorus($symb. P$) into the drainage system. Likewise, the treatment plants in Germany have been considering both phosphorus and nitrogen elimination[1].

Phosphorus compounds in crude, municipal wastewater can be simplified into a dissolved and a particulate state. In their respective states, the compounds are organic or inorganic[2].

By flocculation particulate phosphorus can be separated by sedimentation. The amount of phosphorus released by humans into the drainage system is around 1.8 grams per inhabitant per day[3].

The German Association for Water, Wastewater and Waste e-V (DWA) explains biological elimination with the help of bacteria and chemical-physical elimination options such as the pre-simultaneous and post-precipitation of the element phosphorus.

Key Words

Elimination, Phosphorus, Sewage Treatment, DWA

1 Introduction

Phosphorus is a vital substance for humans, animals and plants. The element is one of the basic building blocks of energy metabolism and cell growth. However, phosphorus is not one of the substitutable substances and its occurrence worldwide is limited. Inventories are limited to a few countries, for which it is economically feasible, considering the current technological situation, to reduce phosphate rock. Most of the phosphorus produced is used in agriculture for fertilizer and feed[2].

However, this nutrient also contains potential risks to the environment. Plants are limited in their growth to the substance that is present in the least amount. According to this principle, phosphorus is the substance that is crucial for plants.

Consequently, if there is an increased supply of nutrients, then water ecology is called eutrophication[4]. Here, an increased growth of algae and water plants takes place. Subsequently, more oxygen is consumed by the increased biomass. In the medium term, anaerobic conditions set in, in which the plants die, which in turn consumes oxygen. This cycle worsened the quality of the water enormously and is colloquially referred to as "tipping over the water body"[2].

Despite all the efforts made in the past, phosphorus emissions from sewage treatment plants into our waters are still high. In some federal states of Germany, therefore, the limit values for phosphorus in the wastewater treatment plant have recently been significantly reduced close to zero, so that in the medium term, further intensification must also be expected in Saxony-Anhalt.

In order to prevent the eutrophication and to comply official orders, there are various ways in wastewater technology to remove phosphorus from the wastewater.

1.1 Biological phosphate elimination

The basic principle of biological P elimination is based on the uptake of dissolved phosphate by microorganisms in order to transiently gain energy.

The increased "eradication" of phosphate in the early sixties was found in activated sludge plants without the addition of precipitants. The process was dependent on the intensity of the aeration[2].

1.2 Precipitation

Precipitation describes a chemical process that converts dissolved, mostly ionic particles in water into a sparingly soluble form(Figure 1). This transfer is triggered by a so-called precipitant. Sub-

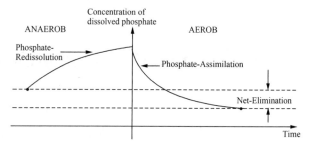

Figure 1　Principal course of the P concentration in a plant for biological phosphorus elimination [Source: DWA Conference: 2017 from Schönberger (modified by translate): 1990;5]

sequently, the resulting product can be physically removed[2-6].

According to DWA 2015, physical processes in wastewater engineering are the following processes, which are briefly addressed:

(1) Sedimentation: Settling of substances contained in the water.

(2) Flocculation: Formation of flocs by transferring the finest substances suspended in the water or distributed colloidally. (colloidal = substances of fine and unrecognizable distribution)

(3) Flotation: Flooding of wastewater ingredients by bubbling fine air bubbles.

(4) Adsorption: Attaching a substance to the surface of another substance.

2　Implementation of the Theory in Practice

2.1　Initial situation

By means of an example of a sewage treatment plant the possible reduction of the parameter phosphorus should be estimated. To this end, the existing phosphate elimination must be explained as well as potential optimization proposals considered and presented.

The biological phosphorus removal reduces the need for precipitant and should therefore continue to be operated.

In order to assess the extent, a long-term series of experiments is needed to study the aerobic and anaerobic phases of the regeneration for their phosphate content. Then one can estimate the degree to which the Bio-P works.

The precipitation of the example wastewater treatment plant is determined manually and continuously promotes the flow. It is a divalent iron chloride applied, which is metered into the return sludge, as simultaneous precipitation. Planned is the enrichment of divalent iron in the return sludge. Subsequently, the precipitant-added return sludge is well mixed in the activation. In the aerobic zone, the precipitant is converted to trivalent iron and can react with the phosphate (Figure 2).

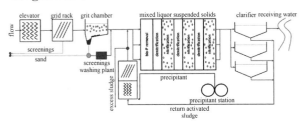

Figure 2　Flow chart (own source: 2017)

2.2　Optimization proposal

One approach to improving the biological phosphate elimination is to control the return sludge ratio, or to deduct via a sludge level measurement or a dry matter measurement in the return sludge. By separating and extracting the biomass from the clear water, the redissolution can be reduced.

The main aim of the optimization concept should be to ensure that the low value of the outlet achieved so far are maintained with a high degree of safety and that they are not exceeded, even with lower monitoring values. It proposes the design of a two-point precipitation.

This project is to be implemented by laying the dosing point with a new tank system from the return sludge to the outlet of the grit chamber. The used divalent iron is maintained as a basic dosage and oxidized by the aeration in the grit chamber. The precipitant manufacturer is in favor of the upgrading of the iron by the grit chamber positive and also the DWA states that a vented grit chamber can cause trivalent iron. Employees of the treatment plant threw in the legitimate doubt whether an oxidation takes place at the end of the aerated grit chamber. There is a need for clarification on this point and requires a series of experiments to provide information. The principle of simultaneous precipitation remains, but it can be ruled out that unreacted iron leaves the process through the excess sludge. Co-precipitated carbon compounds are retained in the system and in the conversion of nitrate during denitrification, but also in biological P elimination. The setting of the basic dosage can, depending on the plant behavior, leave more or less phosphate in the Bio-P.

The precipitating agent station on site is to be converted by a new dosing technology and an automation program for regulated dosing. The planned dosing point is laid by the return sludge to the end of the activation, as simultaneous precipitation. Since the second dosing is to react to fluctuating phosphate levels, the pump technology of the precipitating agent station integrated in an automation program. Optimal would be a phosphate measurement in the manifold structure of the clarifications. The reaction time over the flow path should be sufficient to obtain plausible measurement results and to control the dosing pumps. A link between the measured value and the delivery rate of the pump in the process control system should be established for daily inspection(Figure 3).

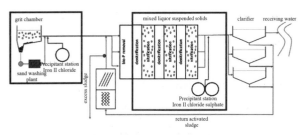

Figure 3　Flow chart with integration of the optimization concept (own source: 2017)

3　Conclusion

On the one hand, the population is obliged to preserve and protect the earth as a habitat for subsequent generations of humans, animals and plants. On the other hand, we are invited to deal with increasing population with the subject of wastewater treatment, as well as the recovery and use of containing substances.

Especially in wastewater treatment, many wastewater treatment plants are under pressure when it comes to phosphorus, but small changes in plant structures can improve elimination and work towards future limits.

References

[1] Schneider, Ralf (2011): Grundlagen für den Betrieb von Belebungsanlagen mit gezielter Stickstoff- und Phosphorelimination. 3. Aufl. Stuttgart: DWA Landesverb. Baden-Württemberg.

[2] Baumann, Peter, Dr., (Hrsg) (2003): Phosphatelimination aus Abwasser. München: Oldenbourg Industrieverlag.

[3] Deutsche Vereinigung für Wasserwirtschaft, Abwasser und Abfall e. V. (Hrsg) (Mai 2011): Arbeitsblatt DWA-A 202 Chemisch-physikalische Verfahren zur Elimination von Phosphor aus Abwasser. Hennef.

[4] Kunst Sabine (1991): Untersuchungen zur biologischen Phosphorelimination im Hinblick auf ihre abwassertechnische Nutzung. Veröffentlichungen des Institutes für Siedlungswasserwirtschaft und Abfalltechnik der Universität Hannover, Heft 77.

[5] Schönberger, R. (1990): "Optimierung der biologischen Phosphorelimination bei der kommunalen Abwasserreinigung". Berichte aus Wassergütewirtschaft und Gesundheitsingenieurwesen, TU München, Nr. 93.

[6] Deutsche Vereinigung für Wasserwirtschaft, Abwasser und Abfall e. V. (Hrsg) (2015): Handbuch für Umwelttechnische Berufe Band 3 // Fachkraft für Abwassertechnik. Druckhaus Köthen.

(7) Flood Protection Measures

Mark Weber

Water Engineering, University of Applied Sciences, Magdeburg-Stendal

Abstract

Structural mistakes of the past and a rising risk of heavy precipitation due to climate change make flood protection measures more and more important. This article aims to take a closer look at engineers' thinking of 200 years ago. In addition, current problems are highlighted and an outlook for the future is given to solve these problems in the best possible way.

Key Words

Flood, Climate Change, Rainfalls

1 Introduction

Heavy rainfalls are very dangerous harassments for the people that are living close to rivers or wetlands. Within hours the water levels can rise so high that rivers can leave their normal river bed and spread out in areas where their appearance will be life-threatening. Activities around the protection of the human population and the infrastructure they are living at, are a very important topic in many countries of the world. Because of the climate change, well working flood managements become more and more important within the 20th century.

2 The Engineer "Tulla"

Around 1809, the Engineer "Johann Gottfried Tulla" had have the opinion that the look of a natural bended river bed is a sign for a retrogressive country. Because of that he convinced the German government to straighten up the big river "Rhine" in a huge project. Additionally, the width of the river was reduced, and the depth of the river bed increased. One of his goals was to increase the agriculture crop land area and the settlement area. His project was an example for other rivers and got copied from many other countries too. Till today the shipping industries can transport their products faster from one place to the other.

3 Consequences

But these advantages led to disadvantages as well. Pretty often the tight river bed doesn't have any emergency flood areas. Flood waves run faster through the river. The reaction time for well working flood activities is much shorter compared with the past. The mistakes of the past can intensify the problems of floods nowadays(Figure 1).

Figure 1 © dpa

To reduce the danger of floods, there are different flood protection activities. This lecture will not only show the effect of traditional buildings like dams and dikes during a flood situation. It will show flood activities that are possible to implement nowadays as well.

4 Solutions

4.1 Rebuilding Dikes

One solution is to rebuild tight river beds. It is important to build dikes further backwards or even resign completely on dikes like in the past. An ecological benefit is that then the population of animals and especially insects is increasing in general. Sometimes it even reaches the same amount as it has been in the past.

4.2 Polder

Polder is very important. With a regulated polder it's possible to cut down the top of a flood wave. It is usually a weir that will be opened in flood situations and so a high number of cubic meters can spread on open land where in the best case nobody and nothing get harmed or damaged. The problem is that most of the times the river is surrounded by agricultural farmland and farmers defend themselves against flooding their land and losing their earnings. To save their income the government should pay compensation payments(Figure 2).

Figure 2 © Frans Lemmens

4.3 Emergency Case Management

Every flood situation is different. Because of that it doesn't exist one plan that fits for all floods. The organisations like firefighters, ambulanceman, policeman and ministries of river managements have to work together and share information as soon as possible. The best way is to create an emergency task force where at least one representative of each organisation collects all information he gets and shares them within the crisis squad.

This management should prevent that some agencies work on the same task or no agency work on a task because they think that the other agency is doing the work.

4.4　Improving Technologies

To measure the amount of rain, weather stations are using rain collectors and sending these information to agencies that are working with these data. To have reliable information the number of collectors should be higher. Heavy rainfall situations are often regional and cannot be detected well enough. Ultrasonic sensors can detect rain events even before the water reaches the ground. The problem is that ultrasonic sensors cannot give reliable information about the amount of rain. A new approach tries to solve that problem. It combines these two technologies and the result should end in a faster detection of heavy rainfalls, a faster reaction time and a better protection of the population[1].

5　Conclusion

Flood protection is an important challenge and because of the climate change it can become even more important. The combination of knowledge and new technologies can be the key to reduce the consequences of heavy and long-term rain periods.

References

[1] Hjalte J. D., Maria L., Arnbjerg-Nielsen K., Mikkelsen P. S., Rygaard M., Efficiency of stormwater control measures for combined sewer retrofitting under varying rain conditions: Quantifying the Three Points Approach (3PA), Environmental Science & Policy, 2016 (63), pp. 19-26.

(8) Long-term Study for the Waste Behaviour of Different Population Groups in Germany

Jonas Thiel, Prof. Dr. Ing. Carsten Cuhls

Department of Water, Environment, Construction and Safety

University of Applied Sciences Magdeburg-Stendal

Abstract

In recent times when natural resources are becoming scarcer, the importance of recycling raw materials increases continuously. This requires a sophisticated waste system. Germany acts as a pioneer and a role model for other countries in this field. However, the success of the German waste disposal system is highly dependent on the population and its willingness to separate waste. For this reason, a long-term study with various population groups was carried out by students of the course "Recycling and Disposal Management" at the University of Applied Sciences Magdeburg-Stendal. The idea of the experiment was to give the students an early insight into the development, execution and analysis of a study. A representative result could not be achieved because of the occurrence of systematic failures during the experiment. Nevertheless, it was possible to determine that living conditions and consumer behaviour certainly have an influence on waste behaviour.

Key Words

Long-term Study, Cycling Raw Materials, Population Groups

1 Introduction

Due to the industry's ever-increasing demand for raw materials and the associated shortage of natural resources, waste recycling is playing an important role. Since waste is often highly calorific, energy recovery plays a major role in addition to recycling. Recycling also helps to protect the environment, as the release of harmful substances is minimised.

In order to ensure a sorted separation of recyclables the German population is encouraged to separate their waste. The separation occurs in paper and cardboard, glass, packaging made of plastics and metal (primarily aluminum and tin plate), biowaste and residual waste. For each fraction a waste container is provided by the German government (or administration) which strongly reduces the effort of recycling due to disposal companies and allows high recycling rates. For instance, the current recycling rate of recovered glass is 75 %[1]. For this purpose, it is separated from foreign substances (metal, plastics, etc.) and coloring, then melted down and finally recast[2].

A similar high recycling rate of 70% is even achieved for paper and cardboard[1]. The recycling process is performed in following way: the waste paper is dissolved in water and various chemicals which causes a dissolution of the printing ink (deinking) and a decomposition of the pulp in fine fibers. In order to produce recycled paper out of the pulp, the cleaned fibers are widely distributed, pressed and dried. This process is significantly more energy and water saving than the production of primary fibers made by wood or cellulose[3].

Packaging waste consisted of plastics or metals is disposed of in the so-called "Yellow Bag" or in the "Yellow Bin". The recycling rate for plastics in total is 60%, for aluminum also 60% and for tin plate 70%. However, only 36% of plastic waste is recycled at all. The rest is used, for example, for energy generation[1]. The waste is separated in sorting plants by various physical processes, like airstream sorting, magnetic separation or near-infrared spectroscopy, according to its kind of material. This results in the following correctly sorted fractions: polyethylene (PE), polypropylene (PP), polyvinyl chloride (PVC), polyethylene terephthalate (PET), composites, aluminum and tin plate. Plastics such as PE, PP or PET are very easy to recycle. They are processed into re-granulate by extrusion or are used directly in injection molding processes. Materials such as composites are more difficult to reuse, because they consist of various components and can therefore only be separated cleanly at great expense. They often end up in waste incineration plants where they are used to generate energy.

The last part of German recycling management is defined as residual waste, that cannot be assigned to the before mentioned fractions. Since 2005 the dispose of untreated waste on landfills was prohibited by the German administration. Thus, residual waste is also used for energy recovery due to incineration. The waste heat generated during combustion is used to generate water vapor, which in turn drives a turbine. The rotary motion of the turbine is transferred to a generator, which finally generates electric current[4]. The ashes and slag produced during combustion are often used in road construction. Only the inert residues are deposited. This ensures that only a small amount of pollutants are released into the environment through waste which results in maximum environmental and human protection.

2 Procedure and Purpose of the Long-term Study

The experiment is carried out by students of the course "Recycling and Disposal Management" at the University of Applied Sciences Magdeburg-Stendal. Relatives and acquaintances of the students serve as test subjects. These are assigned to three different groups: students/apprentices, employed persons and pensioners. Each group is divided once again into "sensitized persons" and "non-sensitized persons". The former believe that they are careful to ensure their proper separation of the waste and the latter pay less attention to the separation. In sum, six groups of individual persons are investigated.

In a period of nine months, the selected test persons should have weighed all incoming household waste and documented the results precisely. A total of 30 households with 57 persons take part in the study. The sensitized workers provide six households with 14 persons, the non-sensitized workers six households with 16 persons, the sensitized pensioners 10 households with 17 persons, the sensitized students/apprentices six households with six persons and the non-sensitized students/apprentices two households with four persons.

The aim of this study is to investigate the behaviour of waste disposal of different age groups with divergent sensitivity in waste separation. It should also be determined to what extent the waste behaviour provides information on the consumption behaviour of the test persons. Furthermore, the students of the course "Recycling and Disposal Management" have the chance to gain first experiences in scientific work and data analysis.

3 Results and Discussion

The results of the previously described study are presented in this section. Table 1 summarizes the results for each investigated group. A further distinction is made between waste weight per person and waste weight per household per month. Unfortunately, no representative data could be collected for the group of non-sensitized pensioners, because of poor attendance. The comparison of the two employed groups does not show any major differences in the masses of packaging and residual waste. In fact, the allegedly sensitized persons generate more waste in the two fractions than the non-sensitized persons. There are clearer differences between the two test groups for the fractions paper and cardboard, glass and biowaste. The paper and cardboard waste produced by sensitized subjects (3305 g/p./m.) is approximately four times higher than produced by non-sensitized subjects (829 g/p./m.). Similar tendencies even occur with glass and biowaste quantities. The high weight of glass could be due to the deliberate avoidance of plastic packaging and the alternative purchase of products packed in glass. The lower quantity of paper and cardboard waste and biowaste in the non-sensitized group may be due to the fact that both fractions are often disposed of in residual waste. Thus, the amount of residual waste would have to be significantly higher for the non-sensitized test subjects than for the sensitized persons. Obviously, logical conclusions, such as the assumption that non-sensitized test persons should produce more residual waste than sensitized subjects, partly do not apply.

One possible explanation is the subdivision of age groups. The test persons should divide themselves into sensitized and non-sensitized. It may be that a rather non-sensitized subject describes itself as sensitized, but in reality, it is not. Hence, in further studies the classification of test groups should be carried out by the organisers. Another explanation is the sensitization of non-sensitized test subjects during the experiment. For further studies, this should be prevented by not changing the waste behaviour of the test persons.

A comparison between the groups of pensioners is not possible. Even though, it is noticeable that the quantities of waste generated by the individual groups of sensitized pensioners are very similar to those of sensitized employees. This is explained by the fact that both groups of test persons exhibit similar consumer behaviour.

The differences are most pronounced among both students/apprentices' groups. While sensitized test persons are very careful to separate the waste, this does not apply for non-sensitized subjects. Paper and cardboard, glass and biowaste are not included at all. For this reason, the residual waste fraction should be significantly larger than the amount of the sensitized test subjects. Nevertheless, the quantities of packaging and residual waste fractions are smaller than for the sensitized groups. This may be explained by the fact that the non-sensitized students or apprentices tend to live with their families and do not spend the weekends in their actual flats. This of course also results in less waste in their households.

A comparison of the age groups shows a gradation between employees as well as pensioners and students/apprentices. In general, the groups of students and apprentices generate less waste than the other two test groups. This can be explained by the different circumstances of life. In most cases the income of younger people, as students or apprentices, is lower than that of employed persons, whereby these groups usually exhibit a smaller financial afford ability and consequential a lower waste disposal.

An explanation for the controversial result could be a variation in lifestyle and consumer behaviour of the different test groups and persons, respectively. This is completely ignored in this study. Especially the difference in waste disposal regarding the consumption of fresh or prefabricated food is really high. In further studies, it is therefore necessary to specific the definition of the terms "sensitized" and "non-sensitized". In favour, the most important fact is the kind of purchasing behaviour of the test persons.

Another failure is the uneven distribution of test persons among the groups. In order to receive a more accurate comparison, it is necessary that each group consists of the same number of households and test persons.

Concluding these systematic failures within the results, the long-term experiment cannot be denoted as a representative study.

However, the students of the course "Recycling and Disposal Management" receive first experiences in the development, execution and analysis of a scientific study.

4 Conclusion

Over a period of nine months, a total of six different population groups (consisting of 30 households with 57 persons) had the task of weighing their waste. The evaluation of the results is only representative to a limited extent, as some failures were made during planning and implementation. These systematic failures influenced the results and made a meaningful analysis difficult.

In order to achieve a satisfactorily result, further studies with improved parameters are necessary.

Acknowledgments

I would like to thank my fellow students for their support in collecting and evaluating the results. Furthermore, I would like to thank Prof. Dr. Carsten Cuhls for accompanying this long-term experiment.

References

[1] DSD - Duales System Holding GmbH & Co. KG, Der Grüne Punkt - Nachhaltigkeitsbericht 2015/2016, p. 20.

[2] Jasna Hamidovic: Industrielle Konzepte zum Altglasrecycling, Peter Lang GmbH, Frankfurt am Main 1997, p. 50-55.

[3] Karl J. Thome-Koziemensky: Recycling von Holz, Zellstoff und Papier, EF-Verlag für Energie- und Umwelttechnik GmbH, Berlin 1987. p. 103-110.

[4] Richard Zahoransky: Energietechnik: Systeme zur Energiewandlung, Springer Vieweg, Wiesbaden 2015.

Table 1 Compilation of the average waste quantities in gram. Explanation: W-Weight, P-Person, H-Household, M-Month

Population group	Packaging		Paper and cardboard		Residual waste		Glass		Biowaste	
	WØper P/M	WØper H/M	WØper P/M	WØper H/M	WØper P/M	WØper H/M	WØper P/M	WØper H/M	WØper P/M	WØper H/M
sen. employed persons	1 635	3 815	3 305	7 711	2 385	5 566	1 527	3 564	3 155	7 361
n.-sen. employed persons	1 242	3 312	829	2 212	1 896	5 055	647	1 725	1 306	3 483
sen. pensioners	1 782	3 030	4 608	7 833	3 238	5 505	1 133	1 926	3 264	5 548
n.-sen. pensioners										
sen. students/apprentices	2 359	2 359	3 045	3 045	4 360	4 360	3 658	3 658	1 767	1 767
n.-sen. students/apprentices	483	967	0	0	2 942	5 884	0	0	0	0

(9) Sorting Analysis and Material Testing of Plastic Waste from the Sea

Lars Tegtmeier, Gilian Gerke, Gunther Weißbach
Department of WUBS, University of Applied Sciences Magdeburg-Stendal

Abstract

Plastic waste contaminates the beaches and accumulates at the bottom of the oceans. Ghost nets drift in the sea and become a deadly trap for whales, fish and seabirds. Our blue planet becomes a dump. Together with fishermen on the North Sea and Baltic Sea, NABU, Germany's Nature and Biodiversity Conservation Union recover waste from the sea and dispose it ashore. But the real challenge starts at this point. Together with the Magdeburg-Stendal University of Applied Sciences they are investigating the degradation processes of the "fished" plastic and looking for a sustainable way of recycling. Over several years they are looking what kind of plastics is collected in the fishermen nets and how the material degrades under the influence of saltwater, sunshine and mechanical load. Is a thermal recovery necessary or can be a qualitative plastic recycling to be achieved?

Key Words

Sorting analysis, Plastic waste

1 Introduction

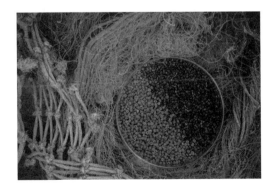

Figure 1 Dolly ropes and Regranulates[1].

Every year an increasing amount of waste ends up in our seas, with a devastating impact on marine flora and fauna. Working together with fishermen at the North and Baltic Seas, NABU, Germany's Nature and Biodiversity Conservation Union, launched the "Fishing for Litter" project in 2011. Main objective to the project is retrieval of marine litter catched by chance during ordinary fishing and environmentally sound disposal onshore. But what happens then with the waste from the sea? In a downstream step, the waste is analysed at Magdeburg-Stendal University of Applied Sciences. With dedicated processing, the way to sustainable recycling is opened up.

2 Sorting Analysis and Goals

Inspired by initiatives in the Netherlands and Great Britain, in 2011 NABU started working with a first local fishing cooperative and other partners to launch the project "Fishing for Litter". The marine waste collected in the nets with the fish catch is collected in big bags on the board and can be disposed without charge by the fishermen in NABU containers at the port. At present around 150 fishermen from 15 German ports are helping to remove the waste from the sea. Fishermen contribute voluntarily; they do not receive financial rewards for their efforts. By the end of 2016, for instance, over 20 t waste had been removed from the sea[2,3]. After being brought on land by the fishermen in the port, the waste undergoes work-intensive manual sorting for material recycling(Figure 2).

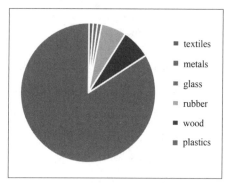

Figure 2 Composition of a sub-sample of 1.26 t fished North Sea waste from 2015 by Magdeburg-Stendal University of Applied Sciences.

The findings obtained can be used for planning future waste avoidance measures. More than 75% of the "fished" waste is made up of products made of plastic. The dominant constituents are films, packaging, net waste and ropes.

Besides the plastic ABS, primarily the mass-produced plastics PP, PE and PVC have been analysed by Magdeburg-Stendal University of Applied Sciences. Another focus was the dominant fraction in mass of net waste made of polyethylene. After the sorting in depth, technical laboratory analysis was performed.

The objective of the following test was to process the collected plastic waste and analyse its material properties. The results provided information on the recyclability of the material and the product options. To this end, samples were subject to a precise material test, in which their behaviour was determined under thermal and physical loading. For the plastics-processing industry, the melting behaviour and the impact strength and tensile strength are important parameters. To determine any degradation of the material caused by its time in the sea, the results of samples were compared with those of new materials and regranulates from the waste from land. Besides knowledge of the material properties, the degree of impurity of the sample material is essential of processing and later recycling. On account of impurities, adhering dirt and mixtures of different material types, in the course of sample preparation, the search was initiated for suitable and proper processing steps to obtain optimum purity. Here it was necessary to compensate for the circumstances typical for marine waste such as salty sands and mineral adhesions. How can the preferred fraction be optimally comminute and which technical challenges must be overcome? Sample preparation forms the basis for the further projects and research work in which marine waste is tested.

3 Material and Sample Preparation

The preferred fraction for this article is dolly ropes. These are thin plastic fibres with a diameter of approx. 1 mm. The individual fibres are intertwined and, as a braid, form a strong and sturdy strand. Tied to the bottom end of the cod-ends, the dolly ropes function as protection against chafing and protect the higher quality main nets against abrasion on the sea bed. Because of the intertwined heterogeneous of the marine litter, the samples had to be manually sorted. To protect the following machines of the abrasive effect of the samples, the material has to be washed. The dolly ropes got an impurity of 18.5%. The high percentage of adhering impurities was caused by the individual threads being interwoven to one mesh, which presents a large surface area for deposits. For better cutting results the samples were melted and crystalline structure formed before.

4 Thermal and Material Testing

With microscopic analyses and a 5 000× magnification by the fraction dolly, ropes an unmistakable degradation of the surface in compare with new reference material could be detected.

With the Fourier-Transform-Infrared Spectrometer (FTIR/ATR) the type of plastic in the samples could be detected. The machine works with light from the infrared spectral range and the material-specific transmissions, reflexion and

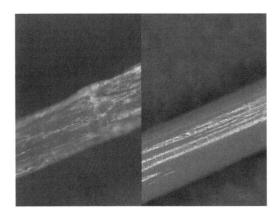

Figure 3 Microscopic analysis of dolly ropes (left: old dolly ropes, right: new reference material) by Magdeburg-Stendal University of Applied Sciences.

ATR reflection is measured. The dolly ropes are based on polyethylene (PE) (Figure 4).

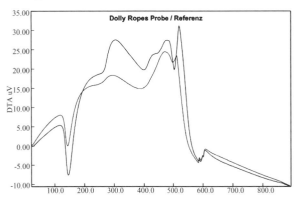

Figure 4 DSC comparison between samples - reference material by Magdeburg-Stendal University of Applied Sciences.

The Differential scanning calorimetry (DSC) compares the results of the FTIR/ATR and gives information about the thermal behaviour. With the heating of the sample, the melting and glass transition temperatures are determined[4]. The parameters are essential for industrial processing and are needed for further analyses of the melt flow index and the preparation of samples.

In the impact strength tests after Charpy, a sample was impacted with a swing hammer. The impact force used was measured and the brittleness and toughness of the material was derived[5]. For the dolly ropes, a considerable loss in quality was detected. While the reference material could be loaded with 166 kJ/m^2, the sample of "fished" material broke already at 45 kJ/m^2. The breaking of the sample was caused by inclusions within the sample and clearly indicated the important role of preparation and cleaning of the material. Even small inclusions e. g. with fine sands which stick to the surface of the individual strands cause weak points within the material.

In the analysis of the melt flow index, the flow ability or viscosity of plastic melts is determined[6]. The melt Mass-Flow Rate (MFR) is an important parameter for the plastics-processing industry. The MFR of the dolly ropes was 0.28 g/10min higher than in the new material and indicated a degradation of the polymer chains and therefore a quality loss.

5 Conclusions

The processing of the plastic waste within the framework of the "Fishing for Litter" project proved a complex problem. Because of the material-specific properties of different fractions, sorting, cleaning and comminution must have very high requirements. These tasks could be only realized so far with a high percentage of manual work. To make material recycling more efficient and economic, other possibilities for optimization of comminution are analysed. Impurities are currently one of the biggest problems and constitute a quality-limiting factor for recycling. Regranulates and products from dolly ropes fraction prove, however, the qualitative realizability of the recycling process.

References

[1] Remiorz K. Picture of the fisher nets. Magdeburg-Stendal University of Applied Sciences, 2017.

[2] Jambeck J R, Geyer R W. Plastic Waste Inputs from Land into the Ocean. In: Science 347/2015, Georgia, USA, 2015, pp. 768-771.

[3] Detloff K. NABU. Together for a clean north and Baltic Sea, 2017.

[4] German Institute for Standardization DIN EN ISO 11357-1. Plastics - Dynamic Differenc Thermal analysis (DSC) - Part 1: General basis. 2010 Berlin, Germany.

[5] German Institute for Standardization DIN EN ISO 179. Plastics - Determination of Charpy - Impact Properties - Part 2: Uninstrumented impact test. 2010 Berlin, Germany.

[6] German Institute for Standardization DIN EN ISO 1133. Plastics - Determination of the melt Mass flow rate (MFR) and the melt flow rate (MVR) of thermoplastics - Part 1: General test procedure. 2012 Berlin, Germany.

(10) Decentralized Rainwater Management

B. Eng. Viktor Plett

Department of Water, Environment, Construction and Safety

Magdeburg-Stendal, University of Applied Sciences

Abstract

The world population of currently about 7.5 billion is projected to increase by 2090 to over 11 billion. Worldwide, the proportion of people living in urban areas is increasing. In 2014, about 53% of the world's population lived in urban areas. Particularly in North and Latin America, this share was particularly high at around 80%, while in Africa and Asia, proportion was still below the average at 40%~46%[1].

A consequence of this combination is that more living space has to be created through a horizontal extension of cities and higher buildings. This results in the sealing of larger contiguous areas on which the natural water balance is greatly altered in the rainfall. The consequences of large-scale sealing are that less rainwater can enter the soil and the urban flood risk increases.

This is where decentralized water management comes in, with the aim of adapting the urban areas more closely to the natural water balance and counteracting the disadvantages of strong sealing.

Key Words

Rainwater, Water management, Resource

1 Introduction

The proportion of the world's population in cities is steadily increasing. In 1950, 71% of the world's population lived in rural areas, while by 2050, around 70% are projected to live in cities, which will be equivalent to an urban population of about 7 billion[2]. This development means that large areas are sealed in order to create living space. This progressive sealing will change the natural water cycle in the urban areas, as shown in Figure 1. The result is less infiltration and evaporation but the share of direct runoff increases. This is particularly detrimental to the recharge of groundwater.

In addition, heavy rainfall events increasingly cause urban flooding.

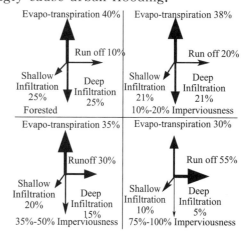

Figure 1 The shift in relative hydrologic flow in increasingly impervious watersheds (Paul and Meyer, 2001)

2 Rainwater Management

Usually the increased rainwater volumes are transported through sewers into rivers and causing high peak flows. In course of the settlement area growth, the sewer systems reach the limits of their capacity faster and faster. For this reason, conventional retention spaces in the form of basins or sewer with storage capacity are created. These measurers only counteract the effects, but not the causes, of the resulting runoff peaks[3].

Figure 2 shows two runoff hydrographs from the same catchment—before and after urban development. The figure demonstrates changes in the runoff hydrograph caused by urbanization[4].

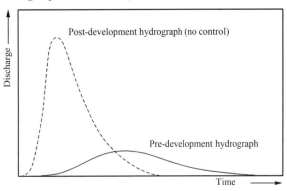

Figure 2 Runoff hydrograph before and after urbanization
(J. Marsalek, 2006)

The difference between the drain hydrographs after the urbanization and before the urbanization is significant. Before the urbanization, the hydrograph spreads over a longer period of time and the runoff intensity is much weaker. Due to the sealing in the urban city, the runoff duration is significantly shorter and more intense among other things, because there is less retention and infiltration.

The measures of the concept of decentralized rainwater management concentrate on the cause of this more extreme and intensive runoff. They begin at the point of origin of the rain runoff and are supposed to delay the runoff. Due to the delay and retention the hydrograph is damped and the runoff is over a longer period of time. The goal is to bring the urban relative water balance back as close as possible to the natural water balance.

The measures of rainwater management are manifold. This is necessary because there are different requirements and conditions in every place in the world. First and foremost, the goal is to create surfaces that are not or less sealed. This automatically leads to changes in the current distribution of the relative hydraulic flow.

The concept of decentralized rainwater management includes a large number of technical and non-technical measures. These measurers can be classified into six components: sage, infiltration, evaporation, retention, treatment and drainage[5].

The proven infiltration methods such as ditches, swales, basins or green roofs rely on these components to achieve the primary objective of these facilities, the avoidance of runoff by infiltration and retention as well as substitution of drinking water to achieve[5].

In order to protect the groundwater, it is necessary to clean the accumulated rainwater from certain surfaces before infiltration. A natural and intensive cleaning takes place in the living soil zone. Technical facilities such as separators and settling tanks must be arranged when heavily polluted rainwater is to be treated.

The measures should already be taken into account during the planning of construction measures, as the worksheet DWA-A 138 and the leaflet DWA-M 153 give recommendations on the handling of rainwater and calculation aids for the construction of systems of infiltration.

In addition to the stated objectives and advantages of decentralized treatment, it should be noted that there are also disadvantages and limitations with this concept. It is always necessary to weigh up if certain measures can be used in urban areas. For example, a road ditch would be ideal for infiltrating the rainfall of the road, but due to space constraints, it is not always possible to build one. In addition, it should be noted that the rainfall may not always be infiltrated without pre-treatment and if additional technical equipment for pre-cleaning is needed, it can be expensive. Furthermore, it should be noted that infiltration is not everywhere the best solution. Geogenic conditions or already existing buildings whose stability can be endangered by the unsealing can limit the infiltration. The stability and functionality of buildings have the highest priority and should not be influenced by the decentralized rainwater management measures.

3 Conclusions

Due to an ever-changing world population, there are constantly changing challenges in the water industry. One of these challenges is to make the natural water balance in urban areas as natural as possible in order to minimize negative consequences for humans and the environment. Decentralized water management will become more important in the future as climate change will increase the occurrence of heavy rainfall events and changes in the urban water balance.

References

[1] DSW, 2014. Datenreport der Stiftung Weltbevölkerung 2014, page 6 et seq. Quoted after de.statista.com: Grad der Urbanisierung nach Kontinenten im Jahr 2014. 2017, https://de.statista.com/statistik/studie/id/12358/dokument/weltbevoelkerung-statista-dossier/, 20.08.2018.

[2] United Nations, 2011. Bunge-Growing world-Annual report 2010, page2- Quoted after de.statista.com: Anteil der in Städten lebenden Bevölkerung weltweit im Zeitraum von 1950 bis 2050. 2017, https://de.statista.com/statistik/studie/id/12358/dokument/weltbevoelkerung-statista-dossier/, 20.08.2018.

[3] Kaiser, Mathias, 2004. Dissertation: Naturnahe Regenwasserbewirtschaftung als Baustein einer nachhaltigen Siedlungsentwicklung mithilfe der Entwicklung und Umsetzung von Modellprojekten, University Dortmund.

[4] J. Marsalek, B. E. Jiménez-Cisneros, P.-A. Malmquist, M. Karamouz, J. Goldenfum and B. Chocat, 2006. Urban water cycle processes and interactions. In: IHP-VI Technical Document in Hydrology N° 78, UNESCO Working Series SC-2006/WS/7, Paris.

[5] Prof. Dr.-Ing. HeikoSieker, 2018. https://www.sieker.de/en/fachinformationen/article/measures-for-decentralized-storm-water-management-245.html, 20.08.2018.

4. Material Science and Engineering

(1)Structures Made of Waterproof Concrete

Franziska Radtke

Department of Civil Engineering,

University of Applied Sciences Magdeburg-Stendal

Abstract

Concrete constructions are widely used in underground structures. The structure's impermeability is generally considered as its weak point. In this paper, the steps of waterproof planning are specified. There are five elements of waterproof planning. The considered elements are the material, force stress in structures, the planning of joints, execution of construction work and building physics.

Planning of structures made of waterproof concrete is grouped in eight steps, which should be followed to achieve a long-lasting waterproof construction.

Key Words

Water Proof Concrete, Durability, Concrete Sealing

1 Introduction

Durability of structures can be reduced by infiltration of water. Sealings should protect elements and structures of adverse effects from water. Infiltrated water could impair the use of this structure in these parts or even block it. Facilities and factory equipment of high value need to be protected of this water.

Sealings could also prevent the running out of liquids from tanks or containers. Contaminations of soil could have disastrous consequences and cause huge costs.

Planning, execution and supervision of sealing construction need to be taken seriously. Often it fails by required conscientiousness.

Sealings need to be safe against the effective water. They cannot lose their protective effect by chemical, mechanical or thermal strain. The choice of the appropriate kind of sealing is conditioned by the type of water, type of foundation soil and the expected strains [1].

One type of sealing is the use of waterproof concrete. Structures made of waterproof concrete are also called white tub. They combine the supporting and proofing functions in one. This is an immense advantage over other waterproofing structures.

2 Waterproof Planning

2.1 Terms and Definitions

Water impermeability

Water impermeability is reached if requirements to restriction and prevention of water flow through concrete joints and construction joints, built-in parts and cracks are achieved.

Design water level

the highest level of groundwater, layer water or high water?

Stress class

Determined by the planner, it considers the type of loading of constructions or structures with moisture or water. Designing water level and ground conditions must be clear for this.

Use class

Determination by the planner, in agreement with client, according to the use of construction.

2.2 Advantages and Disadvantages of Table1 Using Waterproof Concrete for Sealing

Table 1 Advantages and Disadvantages of Using Waterproof Concrete for Sealing

Advantages	Disadvantages
• easier construction • supporting and proofing functions in once • simplified static and constructive arrangement of structure without consideration of permissible surface pressures or failed friction of a sealing skin • reduction of building period • less dependence on the weather because the lost sealing skin • possible leaks are faster to find and easier to fix	• need to be constructed to stay crack free / crack sparse • joints need to get an extra sealing for example by joint tapes • dimensions, formwork and reinforcement need to be coordinated for the concrete • no effect of a vapour barrier

The necessary tightness is secured by using a low water / cement ratio.

Concrete structures whether they are sealed by waterproof concrete or bitumen are showing the same reaction on a long-term basis if concrete with low water / cement ratio is used. In this case, the inside of using site could only appear drying out building moisture and condensation water. Condensation is possible when air with a high moisture content cools off on cold areas. For example, if warm and humid air gets on the inside of a building, it comes to a so-called "beer-glass effect".

Moist inner surfaces are often caused by condensed water and not by infiltrated water. Insulation on the outer surface can normally prevent building of condensation water[1].

2.3 Elements of Waterproof Planning

For building a white tub it needs more than just using waterproof concrete to be in function.

In the process of planning, the following arrangements and single element should be considered.

Material

Concrete with high water penetration resistance

Force stress in structure

crack width and configurations of reinforcements; proof of crack width limitation;

goal: improvement of construction to prevent force

Planning of joints

choice and arrangement of joint sealing

Execution of construction work

compaction, curing

Building physics

thermal insulation, requirements of using, building moisture

The following steps of construction have to be executed for planning of waterproof structures:

1. determination of design water level

(1) Structures Made of Waterproof Concrete

and stress class

2. choice of use class

3. building physical requirements from use

4. determination of minimal wall thickness

5. pressure gradient i and calculative crack width w_k

6. optimize construction concerning force stress

7. determine partition of joints and sealing system

8. built-in parts, penetration

2.3.1 Determination of design water level and stress class

For every planning design water level is an essential basis.

The resulting ground water level, with no lowering of the groundwater level in extremely wet periods, is the decisive design water level.

2.3.2 Choice of use class

Especially requirements of using should get agreement from the client and need to be contractual determined. In the guideline of waterproof concrete this is grouped in use classes.

Use class A

—standard for domestic buildings

—storage space with high quality use

Use class B

—single garages, underground garages

—shaft systems

—storage space with low quality use

2.3.3 Building physical requirements from use

To avoid condensation water on inner surfaces, physical and climatic precautionary measures need to be taken, for example, ventilation, exterior thermal insulation, heating installation and climate control. In use class A, the planner severally has to inform the client about it[2].

2.3.4 Determination of minimal wall thickness

The guideline of waterproof concrete recommends a minimal wall thickness (Table 2). Walls must be thick enough to build in concrete and compact [3].

Table 2 Recommended minimal wall thickness [4]

Stress class		d_{min}[mm] manner of execution		
		In-situ concrete	element walls	pre-fabricated parts
walls	1 pressurized water	240	240	200
	2 soil moisture	200	240 (200)	100
floor slab	1 pressurized water	250	-	200
	2 soil moisture	150	-	100

2.3.5 Define pressure gradient i and calculative crack width w_k

If a waterproof structure with restricted crack formation is chosen, it needs to determine the defined pressure gradient i (Figure 1). This is the quotient of water pressure height and structures thickness at the decisive point.

2.3.6 Optimize construction concerning force stress

Cracks in hardened concrete can accrue if restraint stress in structures is as high as the restraint strength of the concrete. It can happen because of exterior loads like dead loads and live loads or load independent deformations.

The following types of deformations

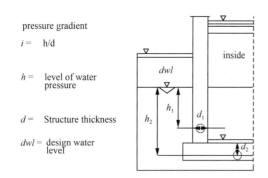

Figure 1　Determination of pressure gradient[4]

must be considered:

- thermal
  - flow of hydration heat
  - temperature change (construction state and use)
  - frost
- hygric (moisture)
  - shrinkage (drying shrinkage)
  - swelling
- uneven settlements or elevations of soil

2.3.7　Determine partition of joints and sealing system

Construction joints on casting segments, planned slip joints aim to reduce force and room joints (settlement joints) between building sections. Type of joint sealing must be coordinated with the existing water pressure. The point where the horizontal and vertical joints collide must be connected as a closed system.

2.3.8　Built in parts, penetration

All penetrations through waterproof structures must be thoroughly planned and sealed. Not only pipe penetrations and line ducts but also anchor sleeves from tensed formwork should be considered. Pipe penetrations should generally go orthogonal through walls or base slabs by the shortest way.

3　Conclusions

Structures made of waterproof concrete represent a special type of construction. Concrete takes the sealing function additional to the supporting function. It creates a simple building construction. This can widely be executed independent from weather conditions. White tubs provide a high level of safety.

For most application ranges structures made of waterproof concrete are standard cases in these days.

The more thoroughly the constructional planning is processed, the more accurate works of execution occurs, the better the expected result is going to be. The result has to be a lasting waterproof construction.

References

[1] Lohmeyer, G.; Ebeling, K., 2009. Weiße Wannen - einfach und sicher. In: Verlag Bau + Technik, Düsseldorf, Germany.

[2] Richter, T., 2008. Bauphysikalische Bewertung von weißen Wannen im Wohnungsbau, beton (58), 2008, H. 4, Verlag Bau + Technik, Düsseldorf, Germany.

[3] DAfstb-Richtlinie "Wasserundurchlässige Betonbauwerke aus Beton (WU-Richtlinie)" (2017-12), Germany.

[4] Zement-Merkblatt H 10: Wasserundurchlässige Betonbauwerke, Bundesverband der Deutschen Zementindustrie e. V., (2012-3), Düsseldorf, Germany.

(2) Research on Resistance to Environmental Corrosion for the Qingdao Bay Bridge Protection

Thomas Plumbohm

Department of WUBS, University of Applied Sciences Magdeburg-Stendal

Abstract

This research work focuses on the comparative study of polyurea's protective properties in contrast to chlorinated rubber coating and epoxy resin coating while being subjected to different environments. Firstly, the general properties of the coatings such as solid content, tack free, curing time, condition of the surface, coating thickness, hardness, tensile strength, elongation at break, impact strength and flexibility at low temperature were studied according to the relevant standards. Secondly, the coated cement mortar samples and control samples were studied by exposure in marine environment, natural exposure cycling, natural exposure cycling under stress, as well as acid and alkali immersion for their impermeability, glossiness, adhesion, chloride/sulfate concentration and corrosion resistance. The results were compared and analysed in terms of their corrosion aging trend, performance and efficiency as a protective material to improve the durability, service life and resistance of the Qingdao Bay Bridge.

Key Words

Resin coating, Concrete corrosion, Durability

1 Introduction

Due to the liberalization of the Chinese market in 1979, China opened up for foreign trade and investment and implanted reforms to initiate a free market[1]. Hand in hand with the rapid economic growth, China is in need of good infrastructure to maintain and ensure further growth. Therefore, China constantly improves its infrastructure and invests large amounts of money in the construction of roads, railway lines, airports, marine infrastructure, etc.[2] But especially on the seaside, corrosion inflicts severe damage to marine constructions and causes great economic losses[3].

Corrosion induces fatal deterioration to metal, alloy and composite fabricated structures causing limited service life and depreciation of infrastructure[4]. Corrosion appears in many different types, and the particularly relevant for the marine industry is the corrosion due to chloride ions and carbon dioxide[5]. Chlorides and carbon dioxide penetrate the concrete and affect the steel without breaking down the concrete first. These two chemicals are most commonly atmospherically born. They will penetrate the concrete without noticeable damage and benefit the corrosion by removing the protective oxide layer of the steel. Corrosion of steel causes cracking, spalling and voluminous corrosion products[6].

Essentially, the most primarily reason resulting in bridges' failures is caused by the combined effects of stress and corrosion. While already corroded, the constant stress by mechanical force fatigues the material additionally. There is no general solution which can be applied due to the high complexity of marine environmental conditions. The research for corrosion protection and long living materials is still not completed, because conventional coatings still fail to ensure a 100-year life expectancy[7]. Consequently, this research work focuses on the comparative study of pure polyurea and other protective coatings in a complex corrosive environment, in order to determine the best coating choice for the Qingdao Bay Bridge and guarantee its long service life.

The Qingdao Bay Bridge, the Qingdao Bay tunnel and the overland expressway are called "Bay Connection Project" with an aggregated length of 41.58 km. The Bay connection is the world's first and longest sea crossing bridge project. The Qingdao Bay Bridge is 26.7 km long and located in the cold sea area of northern part of China. Qingdao city lies in the monsoon climate zone with an average temperature of 12.7 ℃. In extreme conditions the temperature can reach up to 38.7 ℃ and down to −14.3 ℃. Additionally to the wind, waves and freeze, the salinity of the ocean environment requires a high durability of the structure. Furthermore, the northern sea ice demands a high resistance to avert fatigue. To ensure the self-imposed 100-year goal of the bridge design life expectancy, special protection methods are in need to be carried out and key parts must be applied[8].

2 Research Content

2.1 General Properties

According to the relevant standards, the general properties including solid content, tack free, curing time, condition of the surface, coating thickness, hardness, tensile strength, elongation at break, impact strength and flexibility at low temperature were tested under standard environmental conditions in this comparative study for epoxy resin, chlorinated rubber and polyurea coating.

2.2 Protective properties

The 3 different coatings are applied on cement mortar samples. Together with control samples they get subjected to:

(1) Marine atmospheric conditions

(2) Immersion in 3.5% NaCl & 10% Na_2SO_4 solution

(3) Natural exposure and 3.5% NaCl cycle

After 0 d, 7 d, 15 d, 30 d, 45 d and 60 d the impermeability, the glossiness and the adhesion of the substrates are tested and EIS (electrochemical impedance spectroscopy) was performed. Furthermore, the concentration of Cl^- and SO_4^{2-} in the cement mortar samples are tested to observe the prevention effect of the coatings.

2.3 Protective Properties under Bending Stress

Epoxy resin, chlorinated rubber and polyurea coatings were applied on cement mortar samples. Control samples and coated samples were subjected to 30% bending stress and shifted between 3.5% NaCl solution immersion and natural insolation every 12 hours.

3 Experimental Results and Analyses

3.1 General Properties

Polyurea with a solid content of 99.88% features the highest value (epoxy resin 71.58%, chlorinated rubber 46.18%). The higher the solid content is, the lesser the amount of evaporated components is. Evaporated components are solvent and usually harmful to the environment.

Furthermore, polyurea has been proved to reach a dry coating surface in 30 seconds and a complete cure after two minutes. In contrast, chlorinated rubber needs 6 times longer to achieve a dry coating surface and 90 minutes to reach a full cure. Epoxy resin requires 60 minutes for a dry surface and 24 hours for a complete cure.

The tensile strength for polyurea amounts to 21.66 MPa (epoxy resin 8.58 MPa, chlorinated rubber 0.59 MPa). Polyurea also features a higher tear strength with 76.54 N/mm (chlorinated rubber 5.78 N/mm, epoxy resin 2.87 N/mm). Due to the results of the elongation at break, polyurea can be stretched up to 408.12% of its originally size before it breaks (chlorinated rubber 111.31%, epoxy resin 0.23%). It is clearly visible that polyurea exhibits outstanding mechanical performances.

Polyurea features the highest hardness among the tested coatings with A92 (epoxy resin A83, chlorinated rubber A71). The hardness of the surface depends on the durometer range (shore A scale). A high shore A is intended for demanding situations in harsh environments, which makes polyurea a great choice for applications.

Additionally, polyurea features the best flexibility at low temperature, and its surface exhibits no cracks or fractures. The worst flexibility conditions at low temperature offer the epoxy resin. The results may be caused by the different ingredients of the coatings.

3.2 Protective Properties

In regard to the impermeability, uncoated control samples soak up the water very quickly in any environment. Therefore, coatings are imminent to protect the concrete from permeating water to ensure corrosion prevention. After immersion in any displayed environment, the impermeability of polyurea is the best. The epoxy resin, followed by the chlorinated rubber coating, offers a slightly worse impermeability but provides good protection in contrast to the blank control samples. Moreover, polyurea performs about 30% better in 10% Na_2SO_4 solution in contrast to 3.5% NaCl solution in terms of impermeability.

After exposure to different environments polyurea tarnishes very little compared to chlorinated rubber and epoxy resin. It features the best aging resistance in all tested environments due to its hydrogen bonding between the hard segments. In contrast to polyurea, the compactness of chlorinated rubber and epoxy resin is reduced and their resistance lowered. The exposure results in tarnishing, cracking and peeling off, subsequently losing its protective function.

According to experimental results, the epoxy resin coating shows the best adhesion, followed by the chlorinated rubber coating and the polyurea coating with the lowest adhesion. With lasting exposure time in any of the simulated environments, the adhesion of the coatings will decrease. Therefore the longer the immersion is, the more water and ions will permeate through the coating. Likewise, the exposure to UV radiation causes shrinkage resulting in cracks, which further contributes to the decrease in adhesion and finally in the detachment of the coating.

In general, all tested coated specimens exhibit a lower diffusion coefficient in contrast to the blank cement mortar samples. Therefore, the coating hinders the permeating ions, resulting in a better resistance. Especially polyurea shows a favourable permeation resistance. In the marine underwater area, the concrete is in a steady state of immersion. The transfer of chloride and sulfate ions is mainly based on the free diffusion supplemented by pressure infiltration. According to the fitting curves results, the performance of concrete is increasing with prolonging exposure time. While immersed, the concrete still cures. Its compactness grows, resulting in an increased resistance. With extending exposure, the diffusion coefficient will decrease further, due to the ion saturation. However, the corrosion products formed on the surface of the concrete can expand and damage the concrete resulting in cracks. In addition, the dry wet alternation causes concrete salt crystal corrosion and severely damages the concrete.

Via EIS, the electrochemical impedance spectrum of polyurea, epoxy resin and chlorinated rubber is measured in different simulated environments. The research results show that the polyurea coating has a good corrosion resistance and its performance is better than conventional anti-corrosion coatings. 3.5% NaCl aging has little effect on the surface and bulk properties of the polyurea coating. The high molecular chain reaction force is strong and the structure of the coating is still tight.

3.3 Protective Properties under Bending Stress

Giving the experimental results, the additional 30% bending stress is not the major cause for the loss of impermeability. It just slightly increases the permeation ratio of water by 0.05% in contrast to the natural exposure cycle between 3.5% NaCl solution immersion and natural insolation. However, the extra stress causes a significant drop of adhesion (0.57 MPa after 60 days) between the coating and the surface of the concrete.

In all corrosive environments, the diffusion coefficient is decreasing with increasing time. The corrosion resistance of polyurea cycling under stress is the lowest among the tested environments. The additional stress trigger accelerated aging. In compliance with the results of the concentration determinations and diffusion coefficients, the EIS outcomes indicate the negative affect of the additional stress on the corrosion resistance, due to the much smaller capacitance arc radius.

4 Conclusions

Polyurea is doubtlessly the best choice for the harsh marine environment at the seaside of Qingdao and should be taken into consideration for the "Qingdao Bay Bridge". It offers an optimum performance for long term usage, featuring rapid curing, excellent flexibility, high durability, moisture insensitivity, low permeability and little volatile organic compounds (VOC's). Unlike conventional coatings it retains most physical properties over prolonged time.

References

[1] Morrison W M. China's economic rise: history, trends, challenges, and implications for the United States[J]. Current Politics and Economics of Northern and Western Asia, 2013, 22(4): 461.

[2] Zhang W, He L, Zhao X. China business guide: the logistic industry [R]. Beijing: China Council for the Promotion of International Trade, 2009.

[3] Koch G H, Brongers M P H, Thompson N G, et al. Corrosion costs and preventive strategies in the United States. Report by cc technologies laboratories, Inc. to federal highway administration (fhwa), office of infrastructure research and development [J]. Office of Infrastructure Research and Development. Report FHWA-RD-01-156, 2001.

[4] Sambyal P, Ruhi G, Dhawan R. Designing of smart coatings of conducting polymer poly (aniline-cophenetidine)/SiO composites for corrosion protection in marine environment [J]. Surface & Coatings Technology, 2015: 3.

[5] Collepardi M. The new concrete [M]. 刘数华,等,译. 北京:中国建材工业出版社, 2008.

[6] Revie RW, et al. Uhlig's Corrosion Handbook [M]. New Jersey: John Wiley & Sons Inc, 2011.

[7] Cafferty E M. Introduction to Corrosion Science [M]. Alexandria: Springer Science + Business Media, 2010.

[8] Meng F, Yang X, Wang L, et al. The first super large cross sea bridge in the north of China—design of Qingdao Bay Bridge[C]// Proceedings of the eighteenth National Conference on Bridge (Volume One), 2008: 77-93.

(3) Preparation and Characterization of Fe_3O_4/P(MAA-EGDMA)/Au Catalyst

Ma Mingliang, Yang Yuying, Su Siqi, Wu Fei, Li Yanzheng, Si Feiyan, Zhang Zhe

School of Civil Engineering, Qingdao University of Technology

Abstract

The functionalized magnetic core shell Fe_3O_4/P(MAA-EGDMA)/Au catalyst with carboxyl group on the surface was successfully prepared and its catalytic performance was studied. The as-prepared materials were characterized by SEM, FTIR, XRD, VSM and UV-vis analysis. Research indicated that the as-prepared Fe_3O_4/P(MAA-EGDMA)/Au catalyst had excellent catalytic activity and exhibited remarkable reusability for the reaction of reducing 4-nitrophenol. In addition, the catalyst retained 80% of its initial catalytic activity after 7 reaction cycles.

Key Words

Magnetic Core-shell Microspheres, Fe_3O_4/P(MAA-EGDMA)/Au, Catalytic Reduction, Recyclable Catalyst

1 Introduction

As the catalyst with high catalytic activity and magnetic recovery, magnetic catalyst has important application value in the field of catalysis[1]. The Fe_3O_4 nanoparticles are often used as carriers for magnetic catalyst due to their uniform particle size, excellent chemical stability, and good dispersion property[2,3]. The magnetic catalyst is prepared by coating different materials on the surface of the Fe_3O_4 particles.

In this work, high magnetic responsive Fe_3O_4 microspheres were prepared by solvent thermal method, and its surface was chemically modified. Then the polymer P(MAA-EGDMA) was coated on the surface of Fe_3O_4 microspheres by distillation precipitation polymerization. So, the magnetic core-shell Fe_3O_4/P(MAA-EGDMA) composite micro-spheres with carboxyl groups were obtained. After that, the Au ions were adsorbed and reduced to Au atoms to obtain Fe_3O_4/P(MAA-EGDMA)/Au catalyst.

2 Experimental Section

2.1 Materials

Ferric chloride hexahydrate ($FeCl_3 \cdot 6H_2O$), sodium acetate trihydrate (NaAc), anhydrous ethanol, ethylene glycol, silane coupling agent KH570, ammonium hydroxide ($NH_3 \cdot H_2O$), acetonitrile (CH_3CN), 2-azobisisobutyronitrile (AIBN), ethylene glycol dimethacrylate (EGDMA), methacrylic acid (MAA), polyethylene glycol (PEG), chloroauric acid ($HAuCl_4 \cdot 4H_2O$), trisodium citrate dihydrate, sodium borohydride ($NaBH_4$), 4-nitrophenol (4-NP).

2.2 Preparation of Fe_3O_4/P(MAA-EGDMA)/Au catalyst

2.5 g of $FeCl_3 \cdot 6H_2O$, 7.2 g of NaAc and 2.0 g of PEG were added to 80 mL of ethylene glycol, and then the mixture was transferred to a stainless-steel reactor. After the mixture was reacted at 200 ℃ for 6 h, the Fe_3O_4 microspheres were obtained.

0.019 g of Fe_3O_4 microspheres were dispersed in a mixed solution containing 64 g of anhydrous ethanol, 18 g of deionized water and 2 g of KH570. Then, 2 mL of $NH_3 \cdot H_2O$ was added and stirred rapidly at 40 ℃ for 24 h to obtain Fe_3O_4 microspheres with reactive double bonds (Fe_3O_4-KH570).

0.13 g of Fe_3O_4-KH570 microspheres, 0.03 g of AIBN, 1 g of MAA and 1 g of EGDMA were added to 80 mL of CH_3CN. After the mixture was reacted at 90 ℃ for 2 h, Fe_3O_4/P(MAA-EGDMA) microspheres were obtained.

0.1 g of Fe_3O_4/P(MAA-EGDMA), 4 mL of 1% $HAuCl_4 \cdot 4H_2O$ and 1 mL of 1% aqueous solution of trisodium citrate dihydrate were added to 20 mL of deionized water, and the mixture was stirred at 50 ℃ for 50 min. Then, 1 mL of 0.5% $NaBH_4$ solution was added dropwise and stirred for 3 h to obtain Fe_3O_4/P(MAA-EGDMA)/Au catalyst.

2.3 Characterization of Fe_3O_4/P(MAA-EGDMA)/Au Catalyst

The morphology and particle size of the samples were observed by scanning electron microscopy (SEM). Figure 1 was an SEM picture of Fe_3O_4/P(MAA-EGDMA)/Au catalyst. It could be observed that the Fe_3O_4/P(MAA-EGDMA)/Au catalyst was spherical and had a good dispersibility. The particle size distribution of the microspheres was approximately 400 nm.

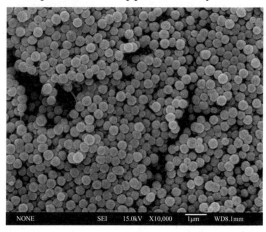

Figure 1 SEM picture of Fe_3O_4/P(MAA-EGDMA)/Au catalyst

Figure 2 was the FTIR spectra of Fe_3O_4/P(MAA-EGDMA)/Au catalyst. The absorption peak at 3 418 cm^{-1} was a stretching vibration absorption peak of a hydroxyl groups (—OH). The absorption peak at 2 972 cm^{-1} was the stretching vibration absorption peak of the C—H bonds. The absorption peak at 2 023 cm^{-1} was an absorption peak of —C≡C≡O. The absorption peak at 1 640 cm^{-1} was a stretching vibration absorption peak of a carbonyl groups (C=O). The absorption peak at 1 102 cm^{-1} was the stretching vibration absorption peak of the —CH_3 bonds. The absorption peak at 869 cm^{-1} was the stretching vibration absorption peak appearing in MAA. The peak at 598 cm^{-1} was the vibration absorption peak of Fe—O. As can be seen from the above, Fe_3O_4/P(MAA-EGDMA)/Au catalyst was successfully synthesized.

Figure 3 was an XRD spectrum of Fe_3O_4/P(MAA-EGDMA)/Au catalyst. The diffraction peaks at 30.1°, 35.6°,

(3) Preparation and Characterization of Fe₃O₄/P(MAA-EGDMA)/Au Catalyst

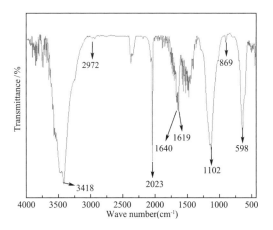

Figure 2　FTIR spectra of Fe₃O₄/P(MAA-EGDMA)/Au catalyst

43.0°, 53.2°, 56.8° and 62.5° corresponded to (220), (311), (400), (422), (511) and (440) crystal faces of Fe₃O₄ microspheres[4]. The diffraction peaks at 38.4°, 44.4° and 64.6° corresponded to the (111), (200) and (311) crystal faces of the gold crystal[5]. It was shown that the Au was successfully loaded on the surface of the magnetic core shell Fe₃O₄/P(MAA-EGDMA).

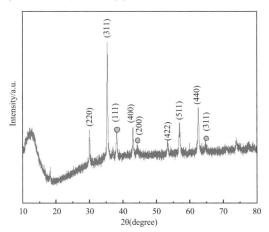

Figure 3　XRD spectrum of Fe₃O₄/P(MAA-EGDMA)/Au catalyst

As the magnetic catalyst, measurement of magnetization saturation was very necessary. As shown in Figure 4, the magnetization saturation value of Fe₃O₄ microspheres was about 80.08 emu/g. The magnetization saturation value of Fe₃O₄/P(MAA-EGDMA)/Au catalyst was 62.16 emu/g after coated with P (MAA-EGDMA) coating. The prepared nanoparticles had good magnetic responsiveness.

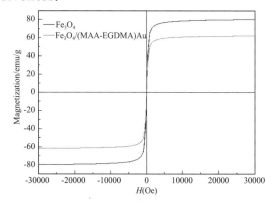

Figure 4　VSM curves of Fe₃O₄ microspheres and Fe₃O₄/P(MAA-EGDMA)/Au catalyst

The catalytic performance of the Fe₃O₄/P(MAA-EGDMA)/Au catalyst was investigated by catalytic degradation of 4-NP as a model reaction (Figure 5). It could be seen from the obtained UV-vis spectra that the intensity of the 4-NP characteristic absorption peak gradually decreased with time, and the absorption peak was hardly visible after 12 minutes. It was shown that the Fe₃O₄/P(MAA-EGDMA)/Au catalyst completely degraded 4-NP in 12 minutes.

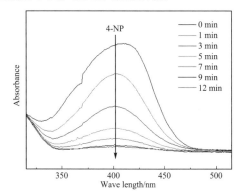

Figure 5　UV-vis spectra of 4-NP and NaBH₄ mixture in the presence of Fe₃O₄/P(MAA-EGDMA)/Au catalyst at different times

Recyclability was an important aspect of the magnetic catalyst. The advantage of the Fe₃O₄/P(MAA-EGDMA)/Au catalyst was that it could be quickly separated from the solution by a magnet and reused in the next cycle. As can be seen from Figure 6, the conversion of the catalyst remained above 80% after 7 cycles of recycling. It could be seen that the catalyst could be efficiently utilized in multiple cycles.

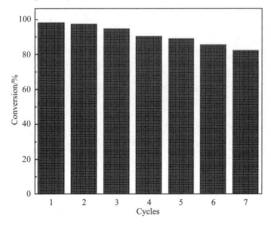

Figure 6 Recycle results of catalytic reactions proceeding in the presence of Fe_3O_4/P(MAA-EGDMA)/Au catalyst

3 Conclusions

(1) In this paper, P(MAA-EGDMA) was coated on the surface of Fe_3O_4 microspheres by distillation precipitation polymerization, then adsorbed Au particles, and Fe_3O_4/P(MAA-EGDMA)/Au catalyst was successfully synthesized. The prepared sample had good dispersibility, uniform particle size of about 400 nm, and high magnetization saturation value of 62.16 emu/g.

(2) The prepared catalyst showed excellent catalytic efficiency (98.2%) in the catalytic reduction 4-NP reaction, and the catalytic efficiency was maintained above 80% after 7 reaction cycles. Fe_3O_4/P(MAA-EGDMA)/Au catalyst can not only degrade 4-NP efficiently, but can also be reused, which had a good application prospect.

References

[1] Chang YC, Chen DH, 2009. Catalytic reduction of 4-nitrophenol by magnetically recoverable Au nanocatalyst. J. Hazard. Mater., 165(1-3): 664-669.

[2] Ma M., Zhang Q., Yin D., et. al., 2012. Preparation of high-magnetization Fe_3O_4-NH_2-Pd(0) catalyst for Heck reaction. Catal. Commun., 17:168-172.

[3] Lotfi Shiri, Mosstafa Kazemi, 2017. Magnetic Fe_3O_4 nanoparticles supported amine: a new, sustainable and environmentally benign catalyst for condensation reactions. Res. Chem. Intermediat., 43(8): 4813-4832.

[4] Ma M., Zhang Q., Dou J., et. al., 2012. Fabrication of 1D Fe_3O_4/P(NIPAM-MBA) thermosensitive-nanochains by magnetic-field-induced-precipitation polymerization. Colloid. Polym. Sci., 290:1207-1213.

[5] Cheng J., Zhao S., Gao W., et. al., 2017. Au/Fe_3O_4@TiO_2 hollow nanospheres as efficient catalysts for the reduction of 4-nitrophenol and photocatalytic degradation of rhodamine B. React. Kinet. Mech. Cat., 121(2): 797-810.

(4) Backward Response Law of Relative Humidity of Unsaturated Concrete Under General Atmosphere Environment in Qingdao Area

Zhao Lixiao, Wang Penggang, Zhao Tiejun, Wang Lanqin, Guo Tengfei

School of Civil Engineering, Qingdao University of Technology

Abstract

It is not as much as the relative humidity of natural environment affects the durability of concrete rather than the internal relative humidity of concrete. The relative humidity of the external environment and the relative humidity of concrete are obtained through embedded sensor. And the backward response of relative humidity of unsaturated concrete in environment with higher humidity is studied. The results show that both the relative humidity of natural environment and the relative humidity in concrete show obvious periodic changes, and the changing trend shows obvious opposite correlation. W/C ratio, concrete depth and environment all have certain influence on backward response of relative humidity.

Key Words

Unsaturated Concrete, Relative Humidity, Backward Response

1 Introduction

The natural environment directly affects the durability of concrete structures[1-3]. Humidity is an important factor affecting shrinkage, carbonation and corrosion rate of steel[4,5]. Many scholars show a lot of researches on the effect of natural environment on the relative humidity of concrete. Jin[6] simulated the capillary water environment, and obtained the different depths of concrete for the 10～40 mm relative humidity evolution law. Ma[7] measured the response curve of temperature and humidity of concrete in Xuzhou with the change of day and night in the natural climate conditions. In fact, the durability of concrete structures is not affected by the relative humidity of the external environment, but by the relative humidity of the concrete itself. The study of the relative humidity response of concrete can reveal the distribution law of relative humidity of concrete and provide a basis for the research of concrete durability.

2 Experiment

The experiment is divided into two different conditions, with shelter and without shelter. After the concrete block is dried to 30% saturation, the block will be placed under the condition of constant temperature(20 ℃) and humidity(50%) for one year. The W/C ratio of the block is 0.2 and 0.5. Wrap 5 sides of the test block with aluminum foil tape, and leave only one test surface. Under shelter conditions, place the specimen in a thermometers creen; it is open around and high above the ground. Under the condition of no shelter, the concrete sample is placed on the white table about 1 m away from the ground and exposed directly to the sun, as shown in Figure 1. The relative humidity of natural environment and the relative humidity response of different depths in concrete are recorded by temperature and humidity sensor.

Figure 1　Testing site

3 Results and Discussion

Figure 2 shows that the change law of relative humidity in different depths of concrete is different from that of natural environment. The relative humidity of natural environment presents obvious periodic fluctuation, whose fluctuation period is about 24 h, and the daily fluctuation amplitude is large. In contrast, the fluctuation of relative humidity in concrete is not significant. The relative humidity value at a certain depth in concrete is basically fixed, and the peak value is opposite compared with that of natural environment. The relative humidity in different depths of concrete varies slightly under the influence of environment, which is mainly reflected in the deeper from the surface layer, the larger of relative humidity value and the smaller of dispersion. This is due to the poor permeability of concrete, and the water vapor will maintain a dynamic equilibrium within a certain range. Water vapor is difficult to exchange for a short time and a long distance, so the concrete interior is less affected by the relative humidity of the surrounding environment.

Figure 3 shows that at the same depth, relative humidity in different exposure condition is significantly different. The relative humidity curve is smoother under the condition with shelter. With the increase of depth, the relative humid-

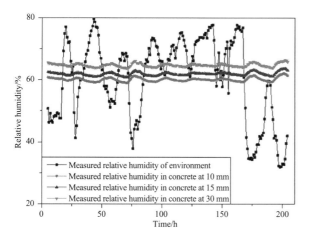

Figure 2　Relative humidity of external environment and concrete of 0.2 W/C with shelter

ity difference between two conditions increased, because the temperature of two different conditions is different and the relative humidity value is affected by the temperature changes.

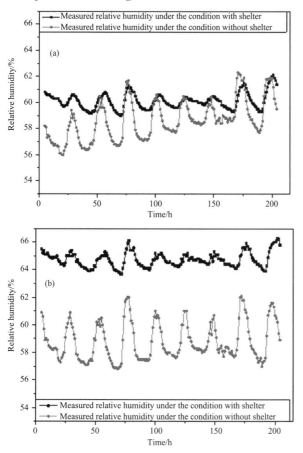

Figure 3　Response of relative humidity in concrete of W/C of 0.2 under the conditions with and without shelter at 10 mm (a) and 30 mm (b)

When without shelter and at different depths, the response of two W/C ratios to relative humidity of concrete is shown in Figure 4. In the case of without shelter, the relative humidity of the block with the same depth of W/C of 0.2 is lower than that with the W/C of 0.5, and the curve is smoother. Because of the drying, the relative humidity in concrete is low. And the porosity of concrete with the W/C of 0.5 is larger, which is easy for water vapor entering from external environment. This allows the moisture within the concrete to be quickly replenished, and achieves water vapor balance at around the relative humidity of 85%. However, the concrete with 0.2 ratio is relatively dense with smaller porosity, and the external water vapor is not easy to enter, so that the internal moisture of the concrete reaches the water vapor balance at about 55%.

As shown in Figure 5, at the same depth, the relative humidity response curve with shelter is smoother. This is because the relative humidity is the relative value affected by temperature. When with shelter, the fluctuation of temperature is relatively small, which is not affected by solar radiation. So the fluctuation of relative humidity is relatively small. Under the same condition, the deeper the concrete is, the stronger the response of the relative humidity inside the concrete is, and this rule is more obvious in the exposed state. This is caused by a combination of internal and external factors. The relative humidity of unsatu-

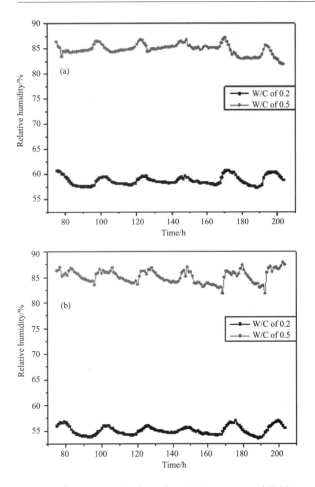

Figure 4 Response of relative humidity in concrete of W/C of 0.2 and 0.5 under the conditions without shelter at 10 mm (a) and 20 mm (b)

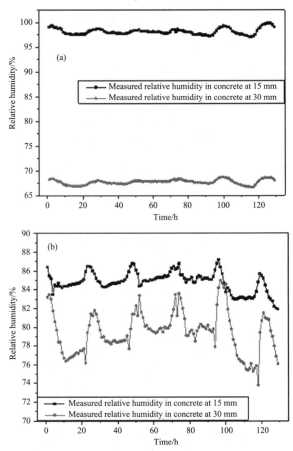

Figure 5 Response of relative humidity in concrete of W/C of 0.5 at 15 mm and 30 mm under the conditions with (a) and without (b) shelter

rated concrete specimens is relatively low. When the external humidity is higher, the internal humidity of concrete increases due to capillary absorption. When the external humidity decreases at a certain depth, on the side facing the external environment, the concrete humidity decreases. On the side facing away from the external environment, due to the lower internal relative humidity, the moisture is still diffused to the interior, resulting in the relative humidity reduction here. The deeper the depth is, the lower the internal relative humidity is. The more serious the water diffusion is, the greater the relative humidity reduction is. As a result, the relative humidity inside the concrete seems to vary more.

4 Conclusions

For unsaturated concrete with low relative humidity, the relative humidity in natural environment fluctuates with time and has remarkable periodicity. There is a strong correlation between the relative humidity of concrete and that of the external environment, and the trend of change between the above two is an opposite correlation. The relative humidity in different depths of concrete varies under the influence of environment, the deeper the depth is, the larger the relative humidity and the smaller the dispersion is. The W/C ratio also has an impact on the relative humidity response of concrete. Under the same depth, the relative humidity of concrete with 0.2 ratio is smaller than that with 0.5. Under the same condition, the deeper the concrete is, the stronger the response of the relative humidity inside the concrete is.

5 Acknowledgements

The authors acknowledge the financial support from the National Natural Science Foundation of China (No. 51608286, 51420105015, 51878254), the National Key & D Program of China (No. 2017YFB0310004-05), the Basic Research Program of China(2015CB655100), and the Source Innovation Program of Qingdao City Grant(NO. 17-1-1-13-jch).

References

[1] Song H W, Kwon S J. Evaluation of chloride penetration in high performance concrete using neural network algorithm and micro pore structure [J]. Cement & Concrete Research, 2009, 39(9):814-824.

[2] Fan XQ. Temperature action and structural design [J]. Journal of Building Materials, 1999, (02): 43-50.

[3] Xu HS, Jiang ZW. Study on autogenous relative humidity change and autogenous shrinkage of high-performance concrete [J]. Journal of Chongqing Jianzhu University, 2004 (02):121-125+136.

[4] Millard S G, Law D, Bungey J H, et al. Cairns. Environmental influences on linear polarisation corrosion rate measurement in reinforced concrete [J]. Ndt& E International, 2001, 34 (34):409-417.

[5] B H, Jang S Y. Effects of material and environmental parameters on chloride penetration profiles in concrete structures[J]. Cement & Concrete Research, 2007, 37(1):47-53.

[6] Jin ZQ, Zhao TJ, Wang BZ, et al. Relative humidity response of concrete under marine environment[J]. Journal of Central South University (Science and Technology), 2014 (10):3608-3613.

[7] Ma WB, Li G. Study on the internal temperature and humidity response of concrete under natural climatic conditions[J]. China Concrete and Cement Products, 2007(2): 18-21.

(5) Molecular Dynamics Study on the Adsorption Properties of Ions at the Surface of Sodium Alumino-silicate Hydrate (NASH) Gel

Hou Dongshuai, Zhang Jinglin

School of Civil Engineering, Qingdao University of Technology

Abstract

The movement of ions in sodium alumino-silicate hydrate gel (NASH) influences the physical and chemical properties of the geopolymer material. The transport of ions in the gel pore fundamentally determines the service life of the geopolymer material. In this paper, molecular dynamics is used to better understand the structure of ions (Cl^-, Na^+, K^+, Cs^+) in the interfacial region of NASH gel. Based on the density distribution of ions on the surface, it is found that the adsorption capacity of cation on NASH gel surface is far stronger than that on chloride ions. And the smaller the ion radius is, the better the adsorption effect is. The adsorption of chloride ions is greatly influenced by cations. It can be seen that the interaction between ions on the nanometer scale and the physical and chemical adsorption of the surface are the key factors that determine ion transport and adsorption.

Key Words

Molecular Dynamics, NASH, Alkali Activated

1 Introduction

New inorganic materials which could increasingly replace conventional cements, plastics and many mineral-based products hold the key to the reduction of world pollution resulting from the manufacturing and use of the older materials.

Geopolymer is a new type of amorphous or semi-crystalline inorganic polymer cementitious material, which has the advantages of convenient production and processing, fast hardening speed, good mechanical properties and excellent durability. Geopolymer has a broad application prospect in municipal, bridge, road, water conservancy, underground, marine and military fields.

However, the corrosion resistance mechanism is still unclear, and it is difficult to explain by experimental means. Therefore, the molecular dynamics method is adopted to explain the micro mechanism from the nanometer scale.

Kalinichev et al. used ClayFF field to simulate different cement hydration products and found their binding capacity to chloride ions[1]. Zhou et al. studied the adsorption and transport properties of chloride ions on C—S—H gel and demonstrated that C—S—H gel with high Ca/Si ratios had better resistance to chloride ion attacking[2]. Hou et al. studied the transport and adsorption characteristics of water molecules and ions in the nano-pores of cement hydration products through molecular simulation, and proved that water molecules near the pore interface showed high density and good orientation, and the interface had certain adsorption capacity for ions[3,4]. In this paper, molecular modeling methods have used to investigate the transport properties of water or ions in the nano-pore channels of cement hydration products.

2 Methodology

2.1 Force field

Lammps software is used to construct the NASH glass model. The ClayFF field chosen for this simulation is established by Cygan et al[5], which is mainly used to calculate the interaction between water molecules, various ions, atoms and the surface of mineral matrix. This force field has been successfully used to simulate the interaction of oxides and hydroxide surfaces with water molecules and ions.

2.2 MD modeling procedure

As shown in Figure 1, the model is built based on the NASH glass model of Si/Al=3. Cationic (Na^+, K^+, Cs^+) solution whose concentration is 0.6mol/L is put in along the direction of the vertical NASH gel matrix. The parameters of Si/Al/Na/O/H, Cs/K/Na/Cl can be found in reference[6].

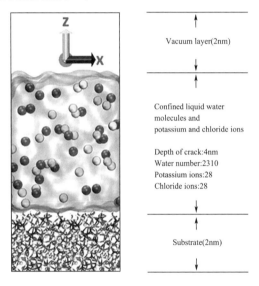

Figure 1　NASH glass interface model diagram of Si/Al=3; Different colored spheres represent different atoms or ions; red is oxygen, green is sodium, yellow is silicon, blue is aluminum, white is hydrogen, light blue is chloride, purple is cations.

The whole simulation is under the NVT ensemble. The system time step is set as 1 fs and the temperature is set as 300 K. The detailed simulation process is as follows: firstly, the NASH substrate is "frozen" by a rigid body method, and only the water molecules and ions move and relax for 1 ns. Secondly, the NASH substrate is released and the whole system is operated for 2 ns to achieve thermodynamic equilibrium; finally, output 3 ns of balance process. The result of molecular dynamics simulation is derived from the analysis of the final equilibrium 3 ns atomic trajectories.

3 Results and Discussion

The density distribution curve of ions can directly reflect the distribution pattern of ions along the direction perpendicular to the NASH gel interface. Then detect the interaction between ions and NASH gel. As illustrated in Figure 2(a), the density curves of cesium ions, potassium ions and sodium ions in the solution at 22Å have obvious strength peaks. This shows that the NASH gel interface has a good adsorption effect on cations. The distribution density of different cations on the surface of NASH gel is in order of size: sodium>potassium>cesium ion. The ionic radius is an important factor affecting ion adsorption. The smaller the ionic radius is, the easier it is to solidify the surface ($r_{Na}<r_K<r_{Cs}$). At the same time, the closer the ions are, the more concentrated the density peaks of sodium and potassium are than cesium. In addition, the smaller the ion radius is, the smaller the hydrated film is formed. That makes it harder for cesium ions with a larger radius to stick to the interface. The broadening density peak of the cesium ion in the figure demonstrates this. It can be seen from schematic diagram Figure 2(b)(c), there is a lot of oxygen of the structure at the interface which will attract a variety of cations to form a group bold, thereby achieving the adsorption effect.

Statistics of the number of ions near

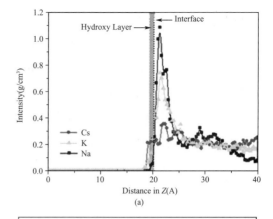

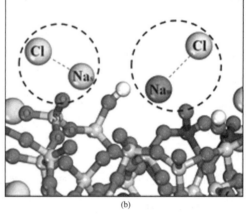

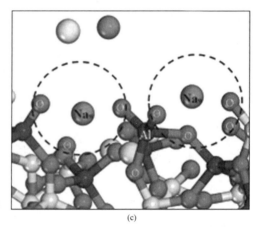

Figure 2 (a)Density distribution of different cations along the Z direction; (b)(c)Schematic diagram of ions and interface effect; Different colored spheres represent different atoms or ions: red is oxygen, green is sodium, yellow is silicon, blue is aluminum, white is hydrogen, light blue is chloride, purple is cations.

the interface during the later balance simulation process can provide a more intuitive understanding of the interface's effect on ion adsorption. As shown in Table 1, the adsorption rate of chloride ions on the interface is 7.1% and the adsorption effect is poor. The adsorption rate of sodium ions, potassium ions and cesium ions on the interface is 30.1%, 22.4% and 17.0% respectively. Therefore, the adsorption effect of the interface on cation is good and that of anions is poor.

Table 1　The adsorption ratio of ions within 5Å at the NASH gel interface

-	Na	K	Cs	Cl
Ratio	30.1%	22.4%	17.0%	7.1%

4　Conclusions

1. The oxygen atoms of NASH surface structure provide a large number of cationic adsorption sites, so the surface of NASH gel is better for cationic adsorption

2. The distribution density of different cations on the surface of NASH gel is arranged in order of size: sodium＞potassium＞cesium ion. This is because the radius of ion is an important factor affecting ion adsorption. The smaller the ion radius is, the more easily it is solidified by the surface ($r_{Na} < r_K < r_{Cs}$). In addition, different ionic hydration radius may affect the adsorption effect.

References

[1] Kalinichev A G, Kirkpatrick R J. 2015. Molecular Dynamics Modeling of Chloride Binding to the Surfaces of Calcium Hydroxide, Hydrated Calcium Aluminate, and Calcium Silicate Phases[J]. Chem. Mater, 14(8): 3539-3549.

[2] Zhou Y, Hou D, Jiang J, et al, 2017. Experimental and molecular dynamics studies on the transport and adsorption of chloride ions in the nano-pores of calcium silicate phase: The influence of calcium to silicate ratios[J]. Microporous Mesoporous Mater, 255, 23-35.

[3] Hou D, Li Z, 2014. Molecular dynamics study of water and ions transport in nano-pore of layered structure: A case study of tobermorite[J]. Microporous Mesoporous Mater, 195(9): 9-20.

[4] Hou D, Li Z, 2013. Molecular dynamics study of water and ions transported during the nano-pore calcium silicate phase: case study of jennite[J]. J. Mater. Civ. Eng, 26 (5): 930-940.

[5] Cygan R T, Liang J-J, Kalinichev A G, 2004. Molecular models of hydroxide, oxyhydroxide, and clay phases and the development of a general force field[J]. The J. Phys. Chem, 108(4): 1255-1266.

[6] Sadat M R, Bringuier S, Asaduzzaman A, et al, 2016. A molecular dynamics study of the role of molecular water on the structure and mechanics of amorphous geopolymer binders [J]. J. Chem. Phys, 145 (13): 1633.

(6) Synthesis and Properties of Phosphorus Flame Retardant Polyurea Elastomers

Li Wenting, Wang Rongzhen, Tong Zhouyu, Song Xinyu, Ma Mingliang, Huang Weibo

Department of Materials Science and Engineering, Qingdao University of Technology

Abstract

Polyurea is an elastomer produced from isocyanate and amino compounds. It has excellent flexibility, aging resistance, medium resistance and good low temperature resistance. However, polyurea is a polymer material, and its research on flame retardant and smoke suppression is also an urgent problem to be solved. In this paper, the formulation of polyurea elastomer was improved, and the polyurea material with flame retardant properties is prepared. The changes in the properties of different NCO contents after synthesizing polyurea elastomers are summarized. The polyurea elastomer is tested for flame retardant and the amount and ratio of the flame retardant are investigated. DMMP and 210 was selected as flame retardants and added to the polyurea formulation. It is found that 200+210 : 600=10 : 9 (the proportion of chain extender 600 in component B is unchanged); 200 : 210=1.41 : 3.59 DMMP accounts for 4% of the total amount of polyurea elastomer material A. This study further enhances the comprehensive performance of polyurea materials and expands its application in civil, construction, military and other fields.

Key Words

Polyurea, Phosphorus Flame, Elastomers

1 Introduction

Although the polyurea material has excellent comprehensive performance, its products such as paints are easy to be incinerated and emit a large amount of toxic gases[1]. Therefore, it is especially important to develop flame retardant polyurea coatings. The focus of research and development is on flame retardants. A flame retardant is a type of physical or chemical mechanism that acts in a solid phase, a liquid phase, or a gas phase (eg, endothermic, covering, chain reaction, etc.) at a specific stage of the combustion process, such as heating, decomposition. A functional additive that inhibits or even interrupts the combustion process during the expansion phase of the combustion or flame, thereby imparting flame retardant to the flammable polymer[2].

In this paper, phosphorus-based flame retardants are used, and the flame retardant effect on oxygen-containing polymer materials is remarkable. Volatile phosphorus compounds are the most effective combustion inhibitors because phosphorus radicals are on average 5 times more effective than bromine radicals and 10 times more effective than chlorine radicals[3,4]. Wang et al.[5] synthesized a new phosphorus-containing flame retardant PFR and added it to epoxy resin.

Phosphorus-containing flame retardants have been widely developed to improve the flame retardant of polymer materials due to their high flame retardant efficiency and low release of toxic gases and smoke[6,7]. In the past ten years, a large number of new organic phosphorus flame retardants have been reported in domestic and foreign literatures, and their flame retardant properties in epoxy, phenolic, polyester, polyolefin, polyamide and other polymer materials have been studied.

In this experiment, two effective phosphorus-based flame retardants are used to modify the polyurea elastomer, and the optimum addition amount and ratio are explored.

2 Experimental Section

2.1 Materials and Equipment

Chain extender 600, Aminopolyether 210, DMMP, Polyether glycol 200 are obtained from Qingdao Shamu New Material Co., Ltd. MDI50 is obtained from Wanhua Chemical Group Co., Ltd.

Electronic universal testing machine MZ-4000D is purchased from Jiangsu Mingzhu Testing Machinery Co., Ltd., oxygen index measuring instrument JF-3 is purchased from Nanjing Jiangning Analytical Instrument Co., Ltd.

2.2 Preparation of Polyurea Elastomer

In this experiment, MDI50 and polyether diol 200 are used for prepolymerization to prepolymers having different NCO contents. The synthesized NCO content is 10%, and the 15% prepolymer is reacted with P-1000 (equivalent 500), chain extender 600 (equivalent 155), 408A (equivalent 187) to prepare a polyurea elastomer.

2.3 Preparation of Flame Retardant Polyurea Elastomer

Since DMMP does not react with the original polyurea elastomer, the addition is physical blending. The calculated DMMP is added to materials A and B, and the mixture is mixed evenly, then A and B are mixed. The 210 and material B are stirred and mixed uniformly, and then react with material A. The stirred polyurea is poured into a mold to obtain a coating film of a certain thickness, and the subsequent mechanical properties and flame retardant properties are tested after curing for a period of time.

2.4 Performance Testing

Tensile performance test is carried out according to GB/T 16777—2008, using dumbbell-shaped specimens; oxygen index test is carried out according to GB/T 2406—1993. The tensile strength is calculated according to formula (1) to the nearest 0.01 MPa.

$$T = \frac{F}{B \times d} \quad (1)$$

T —— tensile strength of the test piece, MPa;

F —— maximum tensile force, N;

B —— the width of the marked line segment of the test piece, mm;

d —— the measured thickness of the test piece, mm.

The elongation at break is calculated according to formula (2) to the nearest 0.01%.

$$E = \frac{L_1 - L}{L} \times 100\% \quad (2)$$

E —— elongation at break of the test piece, %;

L_1 —— the distance between the marking lines when the test piece breaks, mm;

L —— the distance between the starting line of the test piece, mm.

3 Result and Discussion

3.1 Effect of NCO Content on Synthesis of Polyurea Elastomer

Figure 1 has displayed the results of variation curves of tensile strength and elongation at break of polyurea elastomers prepared with different NCO contents. As the NCO content increases, the tensile strength of the polyurea elastomer increases and the elongation at break decreases. Mainly due to the increase of NCO isocyanate groups in the polyurea component A, the reaction amount of the component B is increased, so that the urea groups in the elastomer are increased, and the chemical bonds between the urea groups are tightly combined, resulting in an increase in the stiffness of the polyurea elastomer.

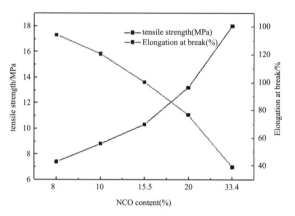

Figure 1 Variation curves of tensile strength and elongation at break of polyurea elastomers prepared with different NCO contents

Due to the increased tensile strength of the polyurea elastomer, the urea bond structure contains two independent hydrogen atoms to form hydrogen bonds between the molecules, enhancing the intermolecular binding force. And the increase in functionality makes the molecular chain change from linear to network structure, which also contributes to the increase of tensile strength.

3.2 Flame Retardant Resistant of Flame Retardant Polyurea Elastomer

Since the effect of 210 and DMMP on the rate of synthesis of polyurea elastomer is different, the gradient design of the mixed addition amount should be opposite to each other.

It can be seen from Figure 2 and Figure 3 that the DMMP addition amount is 4%, and the oxygen index is 27.6%. When the ratio of 210 is 0.4, the flame retardant effect is the best, and the oxygen index is 27.5%. When the flame retardant of 210 and polyether alcohol 200 is added, the oxygen index is 27.8%, which indicates that 210 has a synergistic effect with polyether alcohol 200, and at the same time, it can improve the flame retardant of polyurea elastomer.

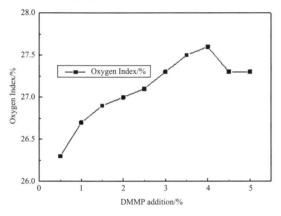

Figure 2 Oxygen index as a function of DMMP addition

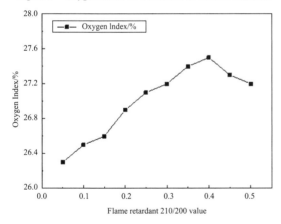

Figure 3 Oxygen index varies with the amount of flame retardant 210 and 200 added

3.3 Thermal Decomposition Behaviour of Phosphorus Flame Retardant Polyurea

It can be seen from Figure 4 that it loses weight quickly at 250 ℃ to 450 ℃, reaching 77.11%, which is the decomposition of polyol and isocyanate. At 110 ℃ to 300 ℃, the polyurea added with 6% DMMP lost 4.53%, and the polyurea added with 4% DMMP lost 2.43%, which is related to the decomposition temperature of DMMP at 180 ℃. At the final temperature, the residual weight of the polyurea added with DMMP is higher than that of the pure polyurea. It is indicated that the phosphorus-containing free radical formed by the decomposition of DMMP reacts with the combustible free radical generated by pyrolysis of the matrix, and also has certain thermal stability.

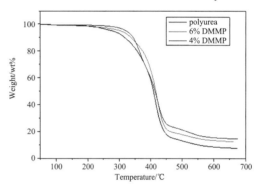

Figure 4 Thermo gravimetric analysis curve of different DMMP contents

4 Conclusions

In this paper, the best formulation for the preparation of slow-reacting polyurea elastomers is found. The effects of prepolymers with different NCO contents on the performance of synthetic polyurea was investigated, and a synthetic prepolymer with an NCO content of 15.5% is finally determined for component A. On the basis of synthesizing polyurea elastomer, a flame retardant is added to increase its flame retardant. A flame retardant polyurea having better flame retardant is synthesized by adjusting the percentage of DMMP and 210 in the polyurea elastomer. Although the mechanical properties of the elastomer decreases slightly, the oxygen index reaches 27.8%.

References

[1] Wen-Wen, W. U., Cui, H. L., Diao, Z. F., Yang, C. J. 2016. Preparation and characterization of flame-retardant spray polyurea coating. Surface Technology, 45(6):22-27.

[2] OuYuxiang. 2008. Handbook of Flame Retardant Plastics. National Defense Industry Press. 1-9.

[3] Babushok, V., Tsang, W. 2000. Inhibitor rankings for alkane combustion. Combustion & Flame, 123(4), 488-506.

[4] Green J. 1992. A Review of Phosphorus-Containing Flame Retardants. Journal of Fire Sciences, 14(6):353-366.

[5] Wang, X., Hu, Y., Song, L., Yang, H., Xing, W., & Lu, H. 2011. Synthesis and characterization of a dopo-substitued organophosphorus oligomer and its application in flame retardant epoxy resins. Progress in Organic Coatings, 71(1), 72-82.

[6] Chen L, Wang Y Z. 2010. A review on flame retardant technology in China. Part I: development of flame retardants. Polymers for Advanced Technologies, 21(1):1-26.

[7] Levchik, S. V., & Weil, E. D. 2006. A review of recent progress in phosphorus-based flame retardants. Journal of Fire Sciences, 24(5), 345-364.

(7) Effect of Curing Regimes on Short and Long-term Compressive Strength Development of Geopolymer Concrete

Cui Yifei[1,2*], Gao KaiKai[1], Obada Kayali[2], Zhao Tiejun[1], Folker H. Wittmann[1]

1. School of Civil Engineering, Qingdao University of Technology, China
2. School of Engineering and Information Technology, University of New South Wales, Australia

Abstract

This study investigated the effect of curing on short and long term compressive strength development of class F fly-ash geopolymer concrete (GPC). With the testing data and micro observation photos, the effect of curing duration and temperature on compressive strength were explored and explained. The macro strength values were measured using compression testing while the micro geo-polymerisation process was observed using scanning electron microscopy (SEM). Experimental data of 3 until 180 days compressive strength of GPC specimens made from three different batches of class F fly-ash (CFA) were recorded and analysed. The micro morphologies of geo-polymer products obtained at different curing durations were studied to identify their degrees of geopolymerisation. The results showed that in contrast to some previous studies, the short and long-term strength gain is related to both heating and ambient curing durations.

Key Words

Geopolymer Concrete, Curing Regimes, Strength

1 Introduction

It is Davidovits, a French chemist who first used the term "geopolymer" to describe synthetic aluminosilicates similar to naturally formed zeolites. The suitable source materials for manufacturing of geopolymers are silica and alumina-rich materials such as metakaolin, milanite, kaolinite and coal combustion fly-ash[1]. Alkaline liquid is the activator for-fly ash based geopolymer and heat is needed for overcoming the activation barrier[2].

There is a burgeoning literature on the examination of appropriate heating temperature and/or duration for class F fly-ash geopolymer composites. Fernandez-Jimenez, Palomo and Lopez-Hombrados[3] place concrete samples into an oven at 85 ℃ for 20 hours after casting. The concrete developes high compressive strength (about 45 MPa) immediately after heat curing. Hardjito and Rangan[4] find that if concrete samples are given an extra 24 hours "resting time" in ambient temperature before heat curing, improvement in the final strength may occur. Hardjito[5] reports an intensive study on compressive strength of GPC cured at a range of temperatures from 30 ℃~120 ℃ for 6~108 hours. Satpute, Wakchaure and Patankar[6] also try heating GPC samples at 30 ℃, 90 ℃ and 120 ℃ for periods of 6, 12, 16, 20 and 24 hours. Their test results show that the increase of compressive strength after 16 hours is not of much significance.

Memon, Nuruddin, Demie and Shafiq find that concrete specimens cure at 70 ℃ produced the highest compressive strength as compared to specimens cured at 60 ℃, 80 ℃ and 90 ℃. Memon et al. also emphasis that the increase in compressive strength after 48 hours is not significant. Mishra et al. further prove that after increasing the curing time from 48 hours to 72 hours, some GPC cubes have even suffered a slight decrease in compressive strength. Van Chanh, Trung and Van Tuan[7] however, point out that when the curing time increases within the range from 24h to 72h, the compressive strength increases gradually. The results of Van Chanh and Van Tuan contradict Khale and Chaudhary's[8] conclusions that geopolymer gains about 70% of its strength in the first 3~4 hours and once it has been cured for 24 hours, its strength keeps constant.

Therefore, this study aims to investigate the correlations between the curing regimes to the short and long-term strength developments of GPC.

2 Experimental Design and Procedures

2.1 Materials Used

2.1.1 Fly-ash

The fly-ash used in this study is ASTM Class F fly-ash from the Eraring thermal power station in New South Wales, Australia. Fly-ash is collected but not produced. So its properties are not stable from batch to batch. Therefore three batches of fly-ash are chosen to uncover possible impact from any abnormal ingredients in the fly-ash collected on one particular date. These batches are collected on three dates: 13-Dec-2012, 31-Jan-2014 and 12-Feb-2014. They are referred to here as 'CFA1', 'CFA2' and 'CFA3' respectively.

2.1.2 Alkaline Liquid

Two alkali solutions; 12 mol NaOH and grade D Na_2SiO_3 are used as the activator of geo-polymerisation. The 12 mol NaOH solution has 361 g NaOH flakes per litre[7], which is prepared by mixing NaOH flakes (>99% purity, supplied by Redox Pyt Ltd., Australia) into deionized water. Grade D Na_2SiO_3 solution contains 29.4 wt% SiO_2 and 14.7 wt% Na_2O.

2.2 Mix Design and Manufacturing

The mix design of GPC is presented in Table 1.

Table 1　　　　Mix design

GPC	14 mm	10 mm	7 mm	Fine (sand)	FA
Mass (kg/m³)	500	310	280	630	420
NaOH	Na_2SiO_3	Water	SP	VM	Total
60	150	31	4	4	2 389

2.3 Specimens and Casting

To allow the alkaline solution to stabilise at room temperature and become homogeneous, the NaOH solution is prepared one day before casting and is stored indoors alongside pre-weighed Na_2SiO_3. During this period, the solution is sealed to prevent any contact with H_2O or CO_2 in the atmosphere.

The concrete samples are cast at different dates and every time only one batch of fly-ash is used to avoid contamination. The samples are separated into 4 different groups (C1, C3, C4, C5) according to their different curing regimes. After casting the samples are moved into an environment-controlled room (ER) at 23±1 ℃ and 50% relative humidity before applying different heat curing regimes. After heat curing samples are moved back to the ER, waiting to be tested on the 3rd, 7th and 28th day.

3 Results and Discussion

3.1 Influence of Heating Duration

Several samples (C3) are heated for 72 hours at 80 ℃ so as to check the effect of heating duration (Figure 1). The results showed, contrary to expectation, that the compressive strength of concrete samples in curing group C3 achieved 23.5%~31.2% increases when compared to the 24 hours heated group C5.

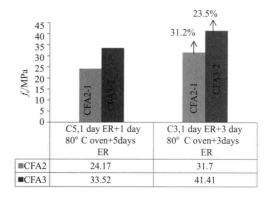

Figure 1 C3 VS C5: Effect of heating duration

3.2 Strength Development after Heating

Another consensus on GPC curing is that the polymerisation process ends during heating, and after more than 24 hours heating, the compressive strength does not vary with the age of concrete[9]. It has been observed that the rate of increase in strength is rapid within 24 hours of heating. Beyond 24 hours, the gain in strength is only moderate. According to these previous publications, it is concluded not only that heat-curing time need not be more than 24 hours, but also that any extended ambient curing after that is not beneficial[9].

Figure 2 shows a comparison of two groups of concrete: C4 has been through 1 day resting at ER, 24h heating at 80 ℃ oven, and it is then tested on the 3rd day after casting. C5 has been stored for 4 days more than C4 in the ER. It is clear that the compressive strength increases during these 4 days.

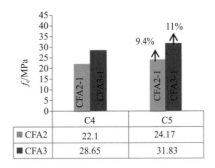

Figure 2 C4 VS C5: Strength development after heating (C4, 1 day ER+1 day 80 ℃ oven+1 day ER)

Results shown in Figure 2 are different from results shown in previous publications, while the strength rise is more distinct as can be seen in Figure 3.

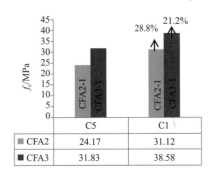

Figure 3 C1 VS C5: Strength development after heating (1 day ER+1 day 80 ℃ oven+26 days ER)

Compared with the 7 day results (C5), the 28 day tests (C1) show more than 20% incremental rise. And, if compared with the 3 day results (C4), about 35% as an average increase is noticed to have been gained after heat curing. It therefore seems that the geo-polymerisation reactions continue beyond the heating period.

3.3 Role of Heat

Figure 4 is an SEM picture for the microporous structure of sample M1 which is cured for 7 days at ER, amplified 1 000 times. The spheroidic fly-ash is dominant in this picture, with hardly any visible geo-polymerisation products.

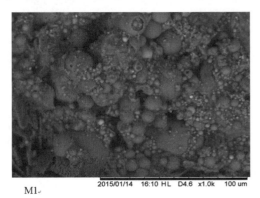

Figure 4 Ambient curing sample M1

After 24 hours of heating, the picture looks very much different. Figure 5 is an SEM picture of sample M3, with 1 000 times magnification. The content is much more complex than in Figure 4. The process of geo-polymerisation has clearly advanced from the stage of dissolution of silica and alumina complexes from the fly-ash into the alkaline (Na^+) medium. A new three dimensional network of silico-aluminates is subsequently formed as can be seen around the fly-ash spheres in Figure 5. From this picture, one main conclusion can be reached and that is temperature is a very important factor for the activation of fly-ash to form a geopolymer binder. This result is consistent with a series of observations conducted by the author.

Figure 5　M3, 1 day ER+1 day 80 ℃ oven

3.4　Long-term Strength Development

Correlations between the age and compressive strength of GPC are shown in Figure 6, in which each bar represents the average compressive strength obtained from three identical samples and the line represents the rates of increases in its strength over time, with its 7-day strength in the zero reference point.

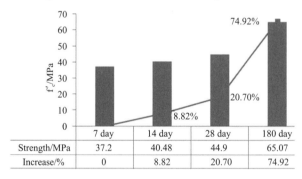

	7 day	14 day	28 day	180 day
Strength/MPa	37.2	40.48	44.9	65.07
Increase/%	0	8.82	20.70	74.92

Figure 6　Rate of strength development after heating

It is clear that the long-term strength increase is significant. The compressive strength of GPC at an age of 180 days increases 74.92% as compared with that after 7 days.

4　Conclusions

(1) There is no doubt that elevated temperature is important in the curing of GPC. It seems that heat curing help markedly in the process of geo-polymerisation and may have specifically assisted the dissolving of fly-ash. Longer heating period has also resulted in higher strength gain as seen in samples heated for 72 hours when compared to those heated for 24 hours. SEM examination shows that samples which are heat treated are able to develop a geopolymer product which is denser when the heating period is extended.

(2) Compressive strength of class F fly-ash geopolymer concrete depends on the treatment history; longer heating time can result in concrete samples gaining higher strength.

(3) In contrast to commonly held perception, strength development after the end of heat curing is observed to be significant in the results of this investigation. Samples of 28 days of age are about 35% stronger than when heating stops after 24 hours.

References

[1] Palomo, A., Grutzeck, M. W., & Blanco, M. T. Alkali-activated Fly-ashes: a Cement for the Future. Cement and Concrete Research, 1999, 29(8), 1323-1329.

[2] Duxson, P., Fernández-Jiménez, A., Provis, J. L., Lukey, G. C., Palomo, A., & Van Deventer, J. S. J.

Geopolymer Technology: the Current State of the Art. Journal of Materials Science, 2007, 42(9), 2917-2933.

[3] Fernandez-Jimenez, A. M., Palomo, A., & Lopez-Hombrados, C. Engineering Properties of Alkali-activated Fly-ash Concrete. ACI Materials Journal, 2006, 103(2).

[4] Hardjito, D. Studies of Fly-ash-based Geopolymer Concrete. Curtin University of Technology, 2005.

[5] Hardjito, D., & Rangan, B. V. Development and Properties of Low-calcium Fly-ash-based Geopolymer Concrete. Perth, Australia: Curtin University of Technology, 2005.

[6] Satpute Manesh, B., Wakchaure Madhukar, R., & Patankar Subhash, V. Effect of Duration and Temperature of Curing on Compressive Strength of Geopolymer Concrete.

[7] Van Chanh, N., Trung, B. D., & Van Tuan, D. Recent Research Geopolymer Concrete. In the 3rd ACF International Conference -ACF/VCA, Vietnam 2008, pp. 11-13.

[8] Khale, D., & Chaudhary, R. Mechanism of Geopolymerisation and Factors Influencing Its Development: a Review. Journal of Materials Science, 2007, 42(3), 729-746.

[9] Rangan, B. V. Fly-ash-based Geopolymer Concrete. Your Building Administrator, 2008, 2.

(8) Review on Research on Epoxy Resin Modified Repair Mortar Based on Alkali-activated Cementitious Materials

Lu Yu, Wan Xiaomei, Zhao Tiejun, Shen Chen
School of Civil Engineering, Qingdao University of Technology

Abstract

Many repair materials, including traditional cement mortar, polymer mortar, and polymer modified mortar have been used to solve the gradual deterioration of concrete. Alkali-activated materials are widely studied for their sustainability and environmental friendliness. The research on characteristics of alkali-activated cementitious materials for repair, as well as the polymer modification mechanism and the repair properties of the modified mortar are reviewed. Compared with ordinary Portland cement, alkali-activated cementitious materials have better mechanical properties and durability, and their unique three-dimensional oxide network structure makes it beneficial for repair. Polymer improves the structural morphology of the slurry hydration product, thereby improving the performance of repair mortar. The addition of epoxy resin to the alkali-activated slag mortar can not only improve the structure of the slurry hydration product by using the polymer, but also solve the problems of high energy consumption, high cost, easy cracking of the mortar, which is beneficial to sustainable utilization.

Key Words

Repair Mortar, Epoxy Resin, Alkali-activated Cementitious Material, Polymer Film

1 Introduction

A large part of the buildings built in China in the early days are in urgent need of reinforcement and maintenance. In the near future, massive constructions in China will be strategically transformed from new built structures to the ones that need renovation. Numerous repair materials, including cement mortar, polymer mortar, and polymer modified mortar have been used to solve this problem.

The appearance of the cement repair mortar is similar to the existing concrete structure. The modulus elasticity of them is close to that of the old concrete matrix, so they have good compatibility. However, they are expected to exhibit higher shrinkage and cracking, especially when cured in a hot dry environment[1, 2]. In the polymer mortar, the polymer used as the binder participates in the curing reaction and can be cured quickly. There are no connected capillary pores inside the material, so they have excellent strength and adhesion to concrete. And both shrinkage and permeability are low[2,3]. However, their thermal compatibility with concrete is poor[3,4], and their high cost limits their wide application.

Polymer modified mortar can modify inorganic materials by adding a certain proportion of polymer, which can make up for the shortcomings of the two. They not only ensure compatibility with cement concrete substrates[3], but also utilize polymer to improve the adhesion and permeability of repair materials[5]. They are considered to be very effective repair materials[6].

Alkali-activated cementitious materials have been extensively studied for their sustainability and environmental friendliness. This sustainable material is a new type of hydraulic inorganic non-metallic glue that can be hydrated after mixing raw materials and activators. They use silicate or aluminosilicate industrial waste (mainly slag and fly ash) as raw materials, using basic compounds or alkaline substances containing industrial waste as activators. They do not require high energy costs in the manufacturing process, and the carbon dioxide produced is reduced by 70% ~ 80%, and greenhouse gas emissions are much lower than ordinary Portland cement[7]. Alkali-activated cementitious materials have the advantages of green, non-polluting and low cost, making them the most promising inorganic non-metallic cementitious materials in this century.

2 Properties of Alkali-activated Cementations Materials

In a strong alkaline solution, the -O-Si-O-Al-O-vitreous structure is rapidly dissolved in the solution to form $[SiO_4]^{4-}$ and $[AlO_4]^{5-}$ tetrahedral units, then a novel three-dimensional network of -O-Si-O-Al-O- bonded materials is obtained by shrinkage and polymerization[8], its unique three-dimensional oxide network structure derived from inorganic polycondensation makes it beneficial for repair[7].

Alkali-activated cementations material is superior to OPC for far lower carbon footprint and less cracking. The main reaction product is a hydrated calcium silicate, like CSH gel. This phase, which has lower C/S ratio, is different from that formed in the Portland cement hydration. Related to the nature and properties of alkali-activated slag cement, it has some advantages over traditional Portland cement, such as higher mechanical strength[9], lower hydration heat[10], better durability under acid attack[11] and freeze-thaw resistance[7]. Huseien[12] believes that the alkali-activated material has a high enough bonding strength, making it suitable as a material for repair engineering.

3 Polymer Modification Mechanism and Repair Mortar Performance

Polymer emulsions for organically

modified repair materials include thermoset emulsions, thermoplastic emulsions, and elastomeric emulsions. The thermosetting emulsion is mainly an epoxy emulsion, the thermoplastic emulsion includes an acrylic emulsion and a vinyl acetate copolymer emulsion, and the elastic emulsion is mainly a styrene-butadiene emulsion (SBR). Most polymer modified mortars exhibit higher compression and flexural strength, better wear resistance, lower permeability and higher bond strength than reference cement mortars[12,13]. The reason is that organic polymers provide a more viscous microstructure and the number of microcracks is reduced[14].

According to the well-known polymer network structure Ohama model[15], the initial polymer particles are deposited on the surface of the gel particles. As the hydration reaction proceeds and the water is continuously consumed, the polymer is gradually confined to the capillary pores. Because the amount of water in the capillary pores is then reduced, the polymer particles flocculate together and form a polymer film on the surface of the hydrated gel.

The mortar slurry structure is denser, and the large pores are significantly reduced, because the larger pores are filled with the polymer. The polymer film can crosslink with the hydration product to form a dense monolith structure[16,17], thereby improving the structural morphology of the slurry hydration product. Ru Wanget et al.[18] found that there are many forms of polymer in the mortar. The polymer particles filled in the larger pores of the mortar and the polymer film bonded to the hydrated product can improve the compactness and moisture of the mortar, and the polymer film formed in the crack of the mortar can prevent further cracking.

Polymer-modified mortars reveal more durability than ordinary repair materials to corrosion resistance, and acid sulphate attack[19]. Epoxy resin mortar can effectively prevent steel corrosion[20], and El-Hawary et al.[21] found that the strength of epoxy emulsion modified concrete in the marine environment still maintained growth. Zhengxian Yang et al.[22] inspected the chloride permeability and microstructure of SBR-modified Portland cement mortars. Electromigration tests demonstrate that the incorporation of SBR latex reduced the general ionic permeability of the mortar. Vinckea et al.[23] demonstrated that styrene-acrylic emulsion(SAE) modification can improve the bio-acid corrosion resistance of concrete.

4 Conclusions

Based on the above superior performance, alkali-activated cementitious materials and epoxy resins can significantly improve the performance of traditional repair mortars. Adding epoxy resin to the alkali-activated mortar can not only make up for the disadvantages of high energy consumption, high cost and easy cracking of ordinary Portland cement, but also improve the bond behaviour and permeability of the repairing material.

References

[1] Mays G C. Durability of concrete structures: investigation, repair, protection. CRC Press, London, 2002, pp. 82-129.

[2] Perkins P H. Repair, protection and waterproofing of concrete structures. Elsevier Applied Science Publishers Ltd., England Google Scholar. 1986.

[3] Fowler D W. Polymer in concrete: a vision for the 21st century. Cem Concr Compos, 1999, 21 (5-6): 449-452.

[4] Saccani A, Magnaghi V, 1999. Durability of epoxy resin-based materials for the repair of damaged cementitious composites. Cem Concr Res, 1999, 29(1): 95-98.

[5] Ye D M, Sun Z P, Zheng B, Feng Z J. Current research status and development initiative of polymer-modified cementitious repair material. MATER REV, 2012, 26 (4):131-135.

[6] Ohama Y. Polymer-based materials for repair and improved durability: Japanese experience. Constr Build Mater, 1996, 10(1):77-82.

[7] Huseien G F, Mirza J, Ismail J, Ghoshal S K, Hussion A A. Geopolymer mortars as sustainable repair material: A comprehensive review. Renew. Sust. Energ. Rev., 2017, 80: 54-74.

[8] Fu Y, Cai L, Wu Y. Freeze-thaw cycle test and damage mechanics models of alkali-activated slag concrete. Constr Build Mater, 2011, 25(7): 3144-3148.

[9] Puertas F, Martínez-Ramírez S, Alonso S, Vázquez T. Alkali-activated fly ash/slag cements: strength behaviour and hydration products. Cem Concr Res, 2000, 30(10): 1625-1632.

[10] Bakharev T, Sanjayan J G, Cheng Y B. Resistance of alkali-activated slag concrete to acid attack. Cem Concr Res, 2003, 33 (10): 1607-1611.

[11] Puertas F, Amat T, Fernández-Jiménez A, Vázquez T. Mechanical and durable behaviour of Z cement mortars reinforced with polypropylene fibres [J]. CemConcr Res,2003,33(12): 2031-2036.

[12] Huseien G F, Mirza J, Ismail M, Ghoshal S, Ariffin MAM. Effect of metakaolin replaced granulated blast furnace slag on fresh and early strength properties of geopolymer mortar. Ain Shams Eng J. 2016.

[13] El-Hawary M M, Abdul-Jaleel A. Durability assessment of epoxy modified concrete [J]. Constr Build Mater, 2010,24(8): 1523-1528.

[14] Colangelo F, Roviello G, Ricciotti L, Ferone C, Cioffi R. Preparation and characterization of new geopolymer-epoxy resin hybrid mortars. Materials, 2013, 6(7): 2989-3006.

[15] Ohama Y. Polymer-based admixtures. Cem Concr Compos, 1998,

20(2-3): 189-212.

[16] Huang Z W, Chen W, Li Q, Wang M, Fan J F. Mechanical properties and microstructure of waterborne epoxy resin modified cement mortar. B CHIN CREAM SOC, 2017, 36(08): 2530-2535.

[17] Knapen E, Van GD. Cement hydration and microstructure formation in the presence of water-soluble polymers. Cem Concr Res, 2009, 39(1): 6-13.

[18] Wang R, Zhang L. Mechanism and durability of repair systems in polymer-modified cement mortars. ADVANCES IN MATERIALS SCIENCE & ENGINEERING, 2015: 1-8.

[19] Hassan K E, Robery P C, Al-Alawi L. Effect of hot-dry curing environment on the intrinsic properties of repair materials. Cem Concr Compos, 2000, 22(6): 453-458.

[20] Thomas C, Lombillo I, Polanco J A, Villagas L, Setien J, Biezma M V. Polymeric and cementitious mortars for the reconstruction of natural stone structures exposed to marine environments. Composites Part B: Engineering, 2010, 41(8): 663-672.

[21] El-Hawary M, Al-Khaiat H, Fereig S. Performance of epoxy-repaired concrete in a marine environment. Cem Concr Res, 2000, 30(2): 259-266.

[22] Yang Z, Shi X, Creighton A T, Creighton, Marijean M, Peterson. Effect of styrene-butadiene rubber latex on the chloride permeability and microstructure of Portland cement mortar. Constr Build Mater, 2009, 23(6): 2283-2290.

[23] Vincke E, Van Wanseele E, Monteny J, Beeldens A, Belie N D, TaerweL, Gement D V, Verstraete W. Influence of polymer addition on biogenic sulfuric acid attack of concrete. International biodeterioration & biodegradation, 2002, 49(4): 283-292.

(9) The Preparation and Properties of a Novel Modified Flame Retardant Polyurea Coating

Li Wenting, Wang Rongzhen, Zhang Lizhi, Tong Zhouyu, Ma Mingliang, Huang Weibo

Department of Materials Science and Engineering, Qingdao University of Technology

Abstract

Polyurea materials are widely used in construction, military, electronic devices and other fields due to their excellent comprehensive properties. However, polyurea is an organic polymer material, which easily decomposes at high temperature and being exposed to fire flames up thus producing a toxic gas which causes poisoning with a lethal outcome. The smoke released by the combustion process reduces visibility, which is not conducive to rescue, so it is important to improve the flame retardant properties of polyurea materials. In this paper, the effects of flake graphite and melamine on the flame retardant of polyurea are investigated. By increasing the amount of flaky graphite up to 1% and the amount of melamine up to 20% and as the oxygen index is 24.9% the flame retardant of polyurea reaches it best performance reaching the national secondary flame retardant standard.

Key Words

Retardant Polyurea, Fire Flames, Flaky Graphite

1 Introduction

Polyurea is widely used as chemical anti-corrosion, pipeline, construction, ship, water conservancy, transportation, machinery and other fields due to its fast setting, solvent-free, quite resistant to a broad range of corrosives and solvents, excellent thermo mechanical properties adhesion and abrasion resistance[1]. However, one of its major defects is its flammability, which limits its further applications. Fan et al[2]. used the manual pressing method to prepare polyurea elastomer and carried out vertical burning test. The experimental results show that the exposed fire with falling objects in the polyurea combustion process will enlarge the flame area and increase the fire hazard. Therefore, it is necessary to improve the flame retardant of polyurea in order to avoid casualties and property losses.

Melamine cyanurate (MCA)[3], a nitrogen containing flame retardant, has received great attention as flame retardant material due to its high nitrogen (N) contents, which is the primary element contributing to the flame retardant of polymer materials. Compared with the traditional halogen flame retardant, it has the characteristics of low smoke density and smoke toxicity, low corrosivity, low flow drop and high flame retardant efficiency. In addition, possessing with a good coloring ability and electrical performance, the ionic impurities aren't find, thus making it environmental-friendly which also corresponds with the trend of the current flame retardant development toward high efficiency and low toxicity. It has drawn a great attention both in domestic and abroad researches in recent years[4].

At present, a large number of studies have shown that a single flame retardant technology and a single flame retardant have a very limited ability to improve the flame retardant of the material, and two or more combinations usually have a very excellent flame retardant effect[5]. The synergistic effect of many flame retardants has become a consensus, such as synergy of nitrogen and phosphorus, synergy of halogens, synergy of phosphorus and halogen, etc.

In this paper, melamine and flake graphite are compounded with flame retardant polyurea. Because scale graphite is easy to agglomerate, a large amount of addition will cause the comprehensive performance of polyurea to decrease. Therefore, it is chosen to add 1wt% of flake graphite [6, 7]

2 Experimental Section

2.1 Materials and Equipment

Polyether diols 200, MDI-50, Q100, 600, Melamine, Flake graphite

Electronic universal testing machine, Thickness gauge, Oxygen index tester.

2.2 Preparation of PU Composites

2.2.1 Preparation of Polyurea

MDI-50 and polyether diols 200 are first premixed as a semi-prepolymer having an NCO content of 15.5 as component A. Then, a mixture of Q100 and 600 in a ratio of 3:2 is used as component B, and finally a component A and a component B are mixed.

2.2.2 Preparation of Flame Retardant Polyurea

Melamine, flake graphite or both are separately added to the above B component, and mixed with the A component to obtain a flame-retardant polyurea material.

3 Result and Discussion

3.1 Performance Comparison Between Polyurea and Flame Retardant Polyurea

It can be seen from Figure 1 that the tensile strength of the pure polyurea elastomer and the flame-retardant polyurea tend to be stable at 5d, and the tensile strength at 30d reaches 33.73 MPa and 20.08 MPa respectively. Compared to the pure polyurea elastomer, the tensile strength of the flame-retardant polyurea elastomer 30d is reduced by 40.47%. At the same time, their elongation at break decrease with the curing time, and the elongation at break of pure polyurea elastomer and flame-retardant polyurea decrease to 266.55% and 250.42% respectively. Compared to the pure polyurea elastomer, the elongation at break of the flame-retardant polyurea elastomer 30d is reduced by 6.05%.

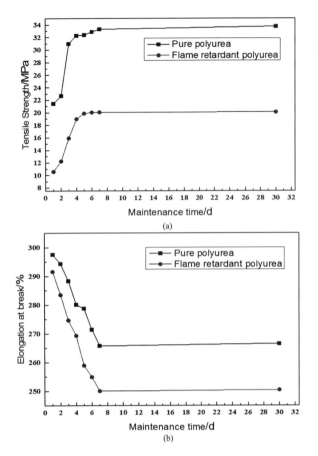

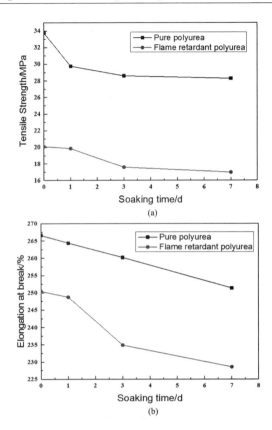

Figure 1 Curve of tensile strength and elongation at break of polyurea with curing time

Figure 2 Curve of tensile strength and elongation at break of polyurea with saturated NaCl immersion time

3.2 Comparison of Soaking Performance of Polyurea and Flame Retardant Polyurea Corrosion Resistant Medium

The polyurea material has good corrosion resistance and can meet the requirements of general construction engineering. This part studies the performance change of corrosion-resistant medium by simulating its service environment, discusses the damage behavior of the coating during service, and provides data foundation and theoretical support for the project.

The performance change of polyurea after immersion in saturated NaCl is shown in Figure 2. After soaking with saturated NaCl solution, the tensile strength and elongation at break of pure polyurea and flame retardant polyurea generally decreased. The decline rates of pure polyurea are 16.16% and 5.70%, respectively. The decline rates of the flame retardant polyurea are 15.44% and 8.72% respectively.

After soaking in saturated NaCl, the tensile strength and elongation at break of pure polyurea and flame retardant polyurea decrease less. The main reason for the change in material properties is that the CO_2 produced by the reaction of the A component isocyanate with the moisture in the air before the reaction of the A component is reacted with the B component ammonia compound during the spray molding. The micropores appear in the material, and during the medium soaking process, the corrosive medium enters the inside of the structure through the micropores, so that the tensile strength thereof is lowered.

In summary, the pure polyurea and the flame-retardant polyurea have good dielectric immersion performance, and the addition of the flame retardant has no negative influence on the medium immersion performance of the polyurea elastomer.

3.3 Effect of Flame Retardant Addition on Flame Retardancy of Polyurea

It can be seen from Table 1 that the LOI value of pure polyurea is only 19.5% at room temperature, which is a flammable polymer material. With the increase of flame retardant, the LOI value increases significantly, reaching 24.9%, an increase of 27.69%, in line with national secondary fire protection requirements.

Table 1　Oxygen index experiment results

	LOI/%
polyurea	19.5
Flame retardant polyurea	24.9

3.4 Thermo Gravimetric Analysis

Evaluation of the thermal stability of polymers is one of the most important indicators for measuring the Fire safety, and thermo gravimetric analysis is the most commonly used method of evaluation. The thermo gravimetric curve of polyurea and its composites under nitrogen is shown in Figure 3.

The pyrolysis process of the flame-retardant polyurea is similar to that of pure polyurea, presenting a two-stage pyrolysis process. The first stage is the pyrolysis of the polyurea backbone and the second stage is the pyrolysis stage of the polyol and isocyanate. The thermal stability of the flame-retardant polyurea material under nitrogen conditions has been improved, and the T_{max} is higher than

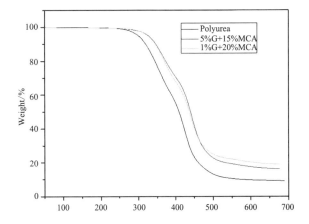

Figure 3　Thermo gravimetric analysis curve

that of pure polyurea. It can be observed that the amount of carbon residue at 500 ℃ has been significantly improved. The flake graphite can exert a physical sheet barrier before it completely degrades itself, slowing the transfer of heat, thereby slowing down the pyrolysis rate of the matrix.

Melamine decomposes between 250 and 450 ℃. By absorbing a large amount of heat, ammonia gas is evolved to form a plurality of polymers, and these substances have a melting point higher than that of melamine itself, and inhibit the progress of the combustion reaction.

4　Conclusions

In summary, we have successfully prepared a flame retardant polyurea by experiment. The mechanical properties of polyurea and flame retardant polyurea are substantially stable within 5 days. After the addition of the flame retardant, the tensile strength and elongation at break are slightly lowered. After immersing in the corrosive salt solution, the tensile strength and elongation at break generally show a downward trend, but the change is not significant. Due to the synergistic effect of flake graphite and melamine, the oxygen index became 24.5%, reaching the national secondary fire protection requirements.

References

[1] T. Mariappan, C. A. Wilkie, Formulation of polyurea with improved flame retardant properties, Journal of Fire Sciences, 31 (2013) 527-540.

[2] Fang Fangqiang, Li Guorong, Zhang Hu, Liao Xixi. Study on the effect of single flame retardant on the flame retardant properties of two-component polyurea elastomers [J]. New Chemical Materials, 2016, 44 (07): 205-207.

[3] W. Tao, J. Li, Melamine cyanurate tailored by base and its multi effects on flame retardancy of polyamide 6, Applied Surface Science, 456 (2018) 751-762.

[4] Jin Dong. Advances in the application of melamine flame retardants [J]. Fine and Specialty Chemicals, 2016, 24(07):26-29.

[5] W. H. Awad, C. A. Wilkie, Further study on the flammability of polyurea: the effect of intumescent coating and additive flame retardants, Polymers for Advanced Technologies, 22 (2011) 1297-1304.

[6] L. C. Zhang, L. Wang, A. Fischer, W. Wu, D. Drummer, Effect of graphite on the flame retardancy and thermal conductivity of P-N flame retarding PA6, Journal of Applied Polymer Science, 135 (2018).

[7] C. Zúñiga, L. Bonnaud, G. Lligadas, J. C. Ronda, M. Galià, V. Cádiz, P. Dubois, Convenient and solventless preparation of pure carbon nanotube/polybenzoxazine nanocomposites with low percolation threshold and improved thermal and fire properties, J. Mater. Chem. A, 2 (2014) 6814-6822.

(10) Preparation and Catalytic Performance of Magnetic Core Shell Fe_3O_4/P(VP-EGDMA)/Ag

Ma Mingliang, Zhang Yue, Wang Ning, Cui Zhonghe, Zhang Kai
School of Civil Engineering, Qingdao University of Technology

Abstract

The functional polymer was coated on the surface of Fe_3O_4 microspheres by precipitation polymerization to obtain magnetic core-shell Fe_3O_4/P(VP-EGDMA) composite microspheres. The Fe_3O_4/P(VP-EGDMA) composite microspheres were further modified to obtain the magnetic core-shell Fe_3O_4/P(VP-EGDMA)/Ag catalyst with surface supported Ag nanoparticles. Catalytic experiments showed that Fe_3O_4/P(VP-EGDMA)/Ag has high catalytic efficiency and excellent recyclability for degradation of methylene blue (MB).

Key Words

Core Shell, Catalytic Performance, Composite Microsphere

1 Introduction

Printing and dyeing wastewater often contains various organic pollutants. MB is a typical organic pollutant in printing and dyeing wastewater. The degradation of MB is crucial for the treatment of printing and dyeing wastewater.

The Fe_3O_4 nanoparticles are super paramagnetic and can be attracted by an external magnetic field and the nanoparticles will disperse when the external magnetic field is removed. In addition, precious metal nanoparticles, especially silver nanoparticles, have good electrical and chemical properties. However, silver nanoparticles have a small surface energy and are easy to agglomerate to reduce catalytic activity. Covering polymer layer on the core surface and then loading the silver ions onto the polymer surface prevents the catalyst from destabilizing due to agglomeration of the silver nanoparticles[1-3].

In this paper, magnetic Fe_3O_4 particles were used as the core, and 4-VP and EGDMA polymers were coated on the outside of Fe_3O_4 particles to synthesize core-shell microspheres. To prepare monodisperse Fe_3O_4 particles, a solvothermal method is employed herein. This step is a chemical reaction using water as a system dispersion medium under high temperature and high pressure conditions[4-6]. What is more, magnetic polymer Fe_3O_4/P(VP-EGDMA) microspheres were prepared by coating polymer layers on magnetic Fe_3O_4[7-9]. Finally, Ag nanoparticles were loaded on the surface of the microspheres to synthesize a recyclable magnetic catalyst. The catalyst after the polymer was separated by an external magnetic field to verify the catalyst performance.

2 Experiment

2.1 Materials

Iron(II) chloride hydrate was purchased in Tianjin BASF Chemical Co., Ltd. Sodium acetate, Ethylene glycol, Poly(ethylene glycol), Acetonitrile, 2, 2'-Azobis(2-methyl propionitrile), Sodium borohydride, Etanol was purchased in Shanghai Aibi Chemical Reagent Co., Ltd. Ethylene dimethacrylate was purchased in Xi'an Lanxiao Technology New Material Co., Ltd. 4-Vinylpyridine was purchased in Shanghai Alpha Biological Reagent Co., Ltd. Ammonium hydroxide was purchased in Yantai Sanhe Chemical Reagent Co., Ltd and silver nitrate was purchased in Tianjin Tianyi Chemical Technology Development Ltd. Methylene Blue trihydrate.

2.2 Preparation of $Fe_3O_4/P(VP-EGDMA)/Ag$

(1) Ferric chloride hydrate ($FeCl_3 \cdot 6H_2O$) was added to the EG, stirred well at 60 °C until completely dissolved, then NaAc and PEG were added and stirring was continued until clear. The reaction was carried out in an autoclave at 200 °C for 6 h. The obtained Fe_3O_4 microspheres were added to a mixed solution containing ethanol and water and stirred rapidly. Add MPS, add $NH_3 \cdot H_2O$ after 2 hours, continue to stir quickly. Magnetic field separation, washing and drying;

(2) The modified Fe_3O_4 microspheres were added to CH_3CN, and then a mixture of AIBN, 4-VP and EGDMA monomers were added, and reacted at 90 °C for 2 h. After the reaction was completed, it was separated by a magnetic field, and washed several times with distilled water and absolute ethanol. Magnetic field separation, washing and drying;

(3) The $AgNO_3$ was sonicated in water, and then the microspheres were added to the above solution for mechanical stirring, followed by the addition of sodium borohydride. Continue to mechanically agitate while maintaining the bath temperature. The magnetic field separation, washing and drying of $Fe_3O_4/P(VP-EGDMA)/Ag$ formation process is shown in Figure 1.

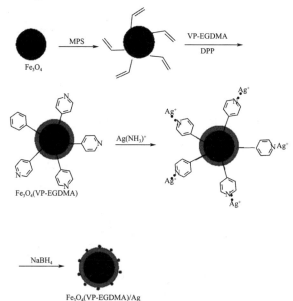

Figure 1　Schematic Illustration of the Synthesis of Fe_3O_4/P (VP-EGDMA)/Ag Microspheres

3 Results and Discussion

Through observation of the Figure 2 (a), it can be found that the obtained Fe_3O_4 is spherical. It can be seen from

Figure 2(b) that there is a thin layer of polymer on the surface of the Fe_3O_4 microspheres. The comparison of a and b shows that the surface of the microsphere is successfully wrapped with a polymer layer.

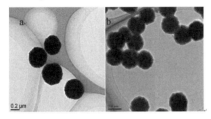

Figure 2　TEM images of (a) Fe_3O_4 (b) Fe_3O_4/P(VP-EGDMA)

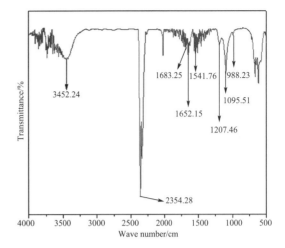

Figure 3　Infrared image of Fe_3O_4/P(VP-EGDMA)

The obtained infrared spectrum of Figure 3 shows an aromatic ring stretching vibration peak of 4-VP, a C-N absorption vibration peak, and an out-of-plane bending vibration peak of aromatic hydrogen. By analysing the structure of 4-VP and EGDMA, it was found that 4-VP and EGDMA were successfully polymerized on the surface of Fe_3O_4 to form Fe_3O_4/P(VP-EGDMA) polymer microspheres.

It can be seen from Figure 4 that the weight decreases with increasing temperature. When the temperature is almost unchanged after 400 ℃, the surface polymer decomposes with increasing temperature, indicating the presence of Fe_3O_4 microsphere surface polymer.

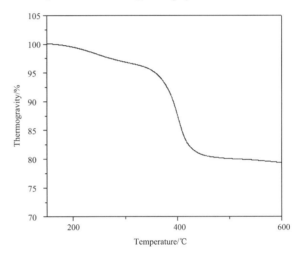

Figure 4　Thermogravimetric analysis of Fe_3O_4/P(V-P-EGDMA)/Ag microspheres

As shown in Figure 5 for the VSM diagram of Fe_3O_4 and Fe_3O_4/P(VP-EGDMA)/Ag, the magnetic saturation strength of Fe_3O_4 microspheres is 80 emu/g, and the magnetic saturation strength of Fe_3O_4/P(VP-EGDMA)/Ag is 60 emu/g. The product still has a high magnetic saturation strength.

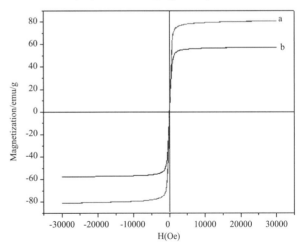

Figure 5　VSM diagram of Fe_3O_4 and Fe_3O_4/P(VP-EGDMA)/Ag

Figure 6 was obtained by XRD characterization of Fe_3O_4/P(VP-EGDMA)/Ag microspheres. Compared with the standard spectrum, the peak position is consistent with the data on the JCPDS card (JCPDSNO. 36-1451), and four distinct diffraction peaks appear at 38.3°, 44.40°, 63.5°, and 77.70° respectively. The four crystal faces of (301), (200), (220), and (311) of the silver indicate that silver nanoparticles have been formed. In addition, the crystal faces marked in blue are (220), (311), (400), (422), (511), and (440) in the anti-spinel structure Fe_3O_4 face-centered cubic lattice. The diffraction peaks corresponding to the crystal planes are 30.1°, 35.5°, 43.1°, 53.5°, 57.1° and 62.8° respectively. It is in good agreement with the standard map JCPDS 75-1609, indicating that the cubic crystal anti-spinel structure of Fe_3O_4 nanocrystals is obtained under the experimental conditions.

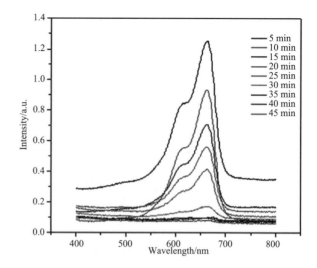

Figure 7　MB concentrate-on changes with time

It can be seen from Figure 8 that the concentration of the MB solution gradually decreases with time and changes significantly at 15 min, which is mainly because the more the catalyst is, the larger the contact area with the MB is, and it is more favorable for MB degradation.

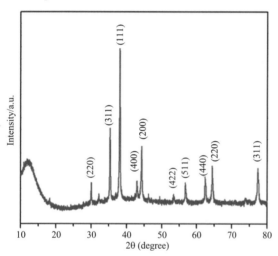

Figure 6　XRD patterns of Fe_3O_4/P (VP-EGDMA)/Ag

It can be seen from the spectrum of Figure 7 that the intensity of the characteristic absorption peak of MB gradually decreases with time, and the absorption peak is hardly visible after 45 minutes. It indicates that the catalyst can completely degrade MB in 45 minutes.

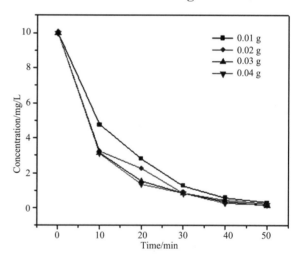

Figure 8　Effect of Catalyst Dosage

At the same time, it can be seen from the Figure 9 that the conversion rate of the catalyst is more than 95% when used for the first time, but when the catalytic effect is reduced to about 85% after two uses, the catalyst is continuously recycled, and the conversion rate is close to 80% after 7 times of use.

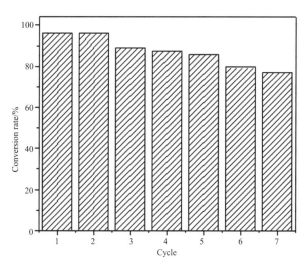

Figure 9 Catalyst Conversion Rate After Multiple Cycles

4 Conclusions

In this paper that focusing on the treatment of MB pollution, a magnetic core-shell Fe_3O_4/P(VP-EGDMA)/Ag catalyst with good catalytic effect and recyclability is proposed. The research content is mainly about two aspects: one is the synthesis of magnetic core-shell catalyst, and the other is to study the catalytic activity of the obtained catalyst. When processing the catalytic reduction of MB, the catalytic time and the amount of catalyst all have an effect on the degradation of MB.

References

[1] Zhang B., Zhang Q., Zhang H., et al. Research progress in preparation and modification of monodisperse nanoparticles. Materials Review, 2009, 23(14): 112-116.

[2] Wang C., Gu H. Preparation and Research Progress of Gold and Silver Nanoparticles Coated Core-shell Microspheres[J]. Journal of Materials Science and Engineering, 2011, 29(6): 958-964.

[3] Wang Z., Zhai S., Zhai B. An Q. "Green" synthesis of magnetic core-shell Fe_3O_4@SN-Ag towards efficient reduction of 4-nitrophenol. Progress in chemistry [J]. 2014, 26(2/3):2-34.

[4] Qin R., Jiang W., Liu H. Preparation and Characterization of Nano Magnetic Ferric Oxide[J]. Journal of Materials Review, 2003, (17): 66-68.

[5] Lv Q., Fang Q., Liu Y. Preparation of ferritic nanoparticles and submicron hollow spheres by hydrothermal method[J]. Journal of Anhui University, 2009, 33(1): 65-69.

[6] Peng D., Bai S., Sha D. Preparation and Characterization of Jadeite Fe_3O_4 by Solvothermal Method[J]. Journal of Synthetic Crystals, 2009, 38(2): 497-501.

[7] Zhao W., Lu T., Zhang H. Progress in Preparation and Application of Core/Shell Structure Organic/Inorganic Composite Microspheres[J]. Material guide, 2009, (7):106-110.

[8] Li G., He T., Li X. Preparation and application of core-shell nanocomposites [J]. Chemical Industry and Engineering Progress, 2011, 23(6): 1081-1089.

[9] Zhao Y., Guo J., Liu Y. Preparation and Application of Core-shell Polymer Microspheres [J]. Shanxi Chemical Industry, 2008, 28(5): 20-23.

(11) Preparation and Properties of Halogen Flame Retardant Polyurea Elastomers

Rongzhen Wang[1], Wenting Li[1], Zhouyu Tong[1], Lizhi Zhang[1],
Xinyue Zhang[1], Mingliang Ma[1], Weibo Huang[1,*]

Department of Materials Science and Engineering, Qingdao University of Technology

Abstract

Polyurea is an elastomer produced from isocyanate and amine compounds. It has very good physical and chemical properties: flexibility, aging resistance, medium resistance, good low temperature resistance and other characteristics. The construction is not susceptible to environmental temperature and humidity and is a green, non-polluting material. In order to further improve the overall performance of polyurea materials and expand their application fields, this article has improved the formulation of polyurea elastomers and prepared polyurea materials with flame retardant properties. In this paper, a polyurea is prepared by reacting a material having an NCO content of 15.5% with a hydroxyl group of a hydroxyl terminated polyether 200 and a MOCA ratio of 7 : 3. The flame retardant is decabromodiphenyl oxide and antimony trioxide, the molar ratio of the two is 3 : 1, and the flame retardant performance is best when the addition amount is 5%.

Key Words

Halogen Flame Retardant, Polyurea, Elastomers

1 Introduction

Polyurea is an elastomeric substance formed by the reaction of an isocyanate component (component A) and an amine compound (component B)[1]. The polyurea elastomer material is prepared by mixing the isocyanate group and the amine compound in a short time by a special spraying device to react the two. The component A prepolymer is reacted with a sealed amine substance, an auxiliary agent, etc. to obtain a liquid compound, which is placed in a natural environment by brushing and spraying, and the terminal amino group of the amine compound is pre-treated by external action. The polymer reacts sufficiently to obtain an elastic material[2]. Polyurea has both high strength and similar elastic modulus to rubber material and high elongation at break.

The Shore hardness of synthetic polyurea elastomer has a relatively large range. It can reach the Shore A30 rubber products, and the highest modulus can reach the Shore D65 anti-impact elastic product (the elastic modulus can reach several hundred MPa), which greatly exceeds the elastic modulus of other rubber products (0.2~10.0 MPa)[3]. Polyurea elastomer has good heat resistance, stability, excellent water permeability resistance, and is resistant to salt spray corrosion, frost resistance, non-toxic, green and environmentally friendly, and has good comprehensive performance. Most polymer materials are flammable and combustible materials themselves, and may emit a large amount of smoke and toxic gases when burned. Polyurea is a polymer material, so there is a potential fire hazard, and flame retardant is very necessary. Flame retardant is a technology that delays or inhibits the propagation of combustion and reduces the probability of occurrence of thermal ignition. It is a technology that fundamentally suppresses and eliminates uncontrolled combustion[4]. There are four flame retardant mechanisms for polymers: condensed phase flame retardant, gas phase flame retardant, synergistic flame retardant and discontinuous heat exchange flame retardant[4,5].

The flame retardant mechanism of the halogen-based flame retardant is as follows. The first step is to prevent the temperature from falling, the second step is to terminate the chain reaction, and the last step is to cut off the heat source[6]. When the lanthanide flame retardant is used alone, the flame retarding effect is minor, and it is usually used together with the halogen flame retardant to form a synergistic flame retardant system and greatly improve the flame retarding effect. The SbOX formed by the reaction of antimony trioxide and hydrogen halide gradually decomposes and releases SbX_3 over a wide range of temperatures and time. They can be used as a diluent to dilute combustible gases. In particular, the thermal decomposition process of SbOX has a strong endothermic effect, which can effectively affect and reduce the decomposition rate of the polymer. Therefore, when a halogen-containing compound or a halogen-containing flame retardant is used in combination with antimony trioxide, an optimum flame retardant effect is usually obtained.

2 Experimental Section

2.1 Materials and Equipment

MDI-50, polyetherdiol 200, hydrochloric acid Diantimony trioxide (Sb_2O_3), MOCA, Bromocresol green, decabromodiphenyl oxide (DBDPO), chlorobenzene, di-n-butylamine thickness gauge, electronic universal testing machine, oxygen index tester.

2.2 Preparation of Polyurea

The MOCA is first dissolved in 200 in a ratio of 3:7 as a component B. A semi-prepolymer having an NCO content of 15.5% is used as the A component. The A component semi-prepolymer is mixed with the same volume of B material, 0.5% of the catalyst is added, and the final product is polyurea.

2.3 Preparation of Flame Retardant Polyurea

From the literature[7], the optimum molar ratio of decabromodiphenyl oxide to antimony trioxide is 3∶1, that is, the optimum mass ratio is 10∶1. Mixing decabromodiphenyl oxide and antimony trioxide in mass ratio (10∶1), adding B component (MOCA dissolved in 220), adding 0.5% catalyst, mixing with equal volume of component A, finally getting the Flame retardant polyurea what we need.

2.4 Performance Testing

Oxygen index is tested according to GB/T 16777-2008. Tensile properties are tested in accordance with GB/T16777-2008.

3 Result and Discussion

3.1 Mechanical Properties of Polyurea

As can be seen from Figure 1, the elongation at break gradually decreases with time and gradually stabilizes after 6 days, while the tensile strength increases with time and then stabilizes. At the beginning, the elongation at break is large and the tensile strength is small, which is due to the urea bond reaction between the B component and the A component. The B component is a polyether diol 200, which is a long-chain structure and has good flexibility. Therefore, the elongation at break is higher at the beginning. As the time elapsed, the polyurea sample was cured at room temperature for a certain period of time, and the urea bond is firmly bonded so that the tensile strength is gradually increased. The elongation at break stabilizes at about 210% within 6 days and the tensile strength stabilized at about 15.50 MPa.

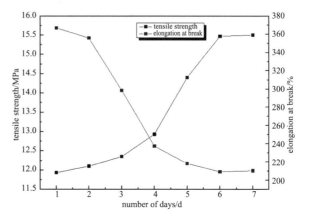

Figure 1 The curves of relation between mechanical properties and cuing time

3.2 Effect of Flame Retardant Addition on Flame Retardancy of Polyurea

Through experiments, when the amount of flame retardant added is 15%, 20% or 25%, it will spontaneously ignite when it leaves the open flame. When the addition amount is 10%, there will be droplets dripping, and after comparison, it is found that the flame retardant performance is better when the addition amount is 1% and 5%. In summary, experiments are carried out using flame retardants at 1%, 3%, 5%, and 8%. The results are shown in Figure 2.

It can be seen from Figure 2 that the oxygen index is 25.80 when no flame retardant is added, and the oxygen index increases after the addition of the flame retardant, and the oxygen index reaches a maximum of 27.4 at 5%. After the addition of the flame retardant, the flame is small at the beginning of the combustion,

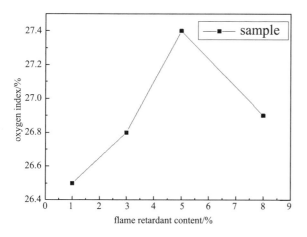

Figure 2 Oxygen Index with Flame Retardant Content Curve

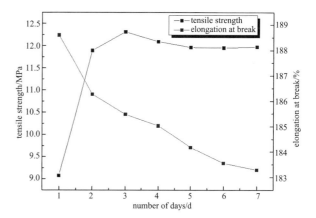

Figure 3 Mechanical properties versus time curve

and then becomes larger. During the combustion process, the sparks are splashed, a small amount of black smoke is emitted, dust is floating, and there is a droplet drop phenomenon.

3.3 Mechanical Properties of Flame Retardant Polyurea

Figure 3 shows the relationship between the mechanical properties of the flame retardant polyurea and the curing time. The elongation at break decreases with time and finally stabilizes at around 183%. After the addition of the flame retardant, the tensile strength begins to rise rapidly, and is basically stabilized at 12 MPa after the third day. Since the added flame retardant is physically blended, the flame retardant is added to the material similarly to the filler, which hinders the combination of urea bonds in the polyurea reaction, and thus the strength is slightly reduced.

3.4 Thermo Gravimetric Analysis

It can be seen from Figure 4 that the weight loss zone of pure polyurea is between 300 ℃ and 450 ℃, and the weight loss rate is 63.69%, which is caused by decomposition of polyol and isocyanate. The weight loss rate of the flame retardant polyurea with a ratio of Sb_2O_3 to DBDPO of 2∶1 is 4.34% at 250 ℃ to 320 ℃, and the weight loss rate of the flame retardant polyurea with a ratio of Sb_2O_3 to DBDPO of 1∶10 is 5.21%. Because Sb_2O_3 promotes the decomposition of the flame retardant, resulting in more weight loss.

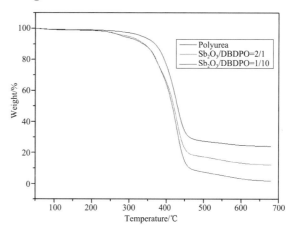

Figure 4 Thermo gravimetric analysis curve

Finally, the results show that the polyurea containing more Sb_2O_3 flame retardant has better flame retardant performance, because Sb_2O_3 is a condensed phase flame retardant, while DBDPO flame retardant is a gas phase.

4　Conclusions

In this article, a flame retardant polyurea is successfully prepared by introducing a halogenated flame retardant with an oxygen index of 27.4. During the 7-day curing process, the tensile strength of the synthesized polyurea gradually increases to 15.5 MPa, and the elongation at break gradually stabilizes at 210%. After adding 5% halogenated flame retardant, since the flame retardant and the polyurea component are physically blended, the tensile strength of the flame retardant polyurea is slightly lowered, but still meets our requirements.

References

[1] Huang Weiwei, Lu Ping. Spraying Polyurea Elastomer Materials [J]. Material guide, 2000, 14(12):33-35.

[2] Sun Zhiheng, Zhang Huiwen. Characteristics, classification and application range of polyurea materials[J]. Water Resources Planning and Design, 2013(10):36-38.

[3] Zwiener C, Sonntag M, Kahl L. Aspartic Acid Esters-A new line of Reactive Diluents for High-Solids Two-Pack Polyurethane coatings In: Proceedings of the ＋wentieth fatipec congress[c]1990. 267-287.

[4] Du BX, Guo ZH, Fang ZP. 2009. Effects of organo-Clay and sodium dodecyl sulfonate intercalated layered double hydroxide on thermal and flame behavior of intumescent flame retarded polypropylene [J]. Polym Degrad Stabil, 94 (11): 1979-1985.

[5] Fang Hailin. Polymer Processing Aids [M]. Beijing: Chemical Industry Press, 2007:85-88.

[6] Tang Ruogu, Huang Zhaoge. Research progress of halogenated flame retardants[J]. Bulletin of Science and Technology, 2012, 28 (1): 129-132.

[7] MarcoZanetti, Giovanni Camino, Domenico Canavese, et al. Fire Retardant Halogen-Antimony-Clay Synergism in Polypropylene Layered Silicate Nanocomposites [J]. Chemistry of Materials, 2002, 14(1):189-193.

(12) Research Progress on Measurement and Calculation Methods of Damping Material Loss Factor

Wu Di

Institute of Functional Materials, Qingdao University of Technology

Abstract

At present, one of the effective methods to control vibration is to use the energy dissipation principle of damping materials. In this paper, several conventional calculation methods, such as cantilever beam resonance method and half power bandwidth method are discussed.

Key Words

Damping Material, Loss Factor, Vibration

1 Introduction

With the progress of science and technology, mechanical equipment tends to be automated and high-speed, vibration and noise problems become more and more serious. Vibration and noise are related to the reliability and stability of the equipment directly. There may be errors or failures in the reading of the equipment, and serious damage to the equipment. In addition, if the vibration problem is not solved for a long time, the fatigue problem of the equipment will also occur, which greatly reduces the service life of the equipment and limits its performance.

Damping technology is one of the most widely used methods to suppress vibration. The technology uses the principle of damping energy dissipation[1].

Damping material is an energy dissipation material that can convert mechanical energy into heat energy. The main index to measure the performance of damping materials is the loss factor of materials. The general test method mainly includes the cantilever beam resonance method, and the calculation method includes the half-power bandwidth method[2,3].

2 Cantilever Beam Resonance Method

2.1 Basic principle

The cantilever beam bending resonance method is a classical method for measuring the dynamic mechanical parameters of damping materials. It is used widely because of its mature theory and convenient implementation.

For the flexural vibration of freely damped structures, the following relationship holds:

$$\beta_2 = \eta \frac{1+eh}{eh} + \frac{1+4eh+6eh^2+4eh^3+e^2h^4}{3+6h+4h^2+2eh^3+e^2h^4}$$

(1)

$$E_2 = E_1 \frac{(u-v)+\sqrt{(u-v)^2-4h^2(1-u)}}{2h^3}$$

(2)

Among them: $e = E_2/E_1$ is the modulus ratio of damping material and steel beam; $h = H_2/H_1$ is the ratio of the thickness of the damping layer to the base thickness; $D = \rho_2/\rho_1$ is the ratio of material density of damping layer to base material density; $u = (1+Dh)(f_{2n}/f_{1n})^2$; $v = 4+6h+4h^2$; f_{1n}, f_{2n} is the Nth order resonant frequency of pure steel beam and composite beam respectively; η is the loss factor of composite beams; β_1 is the loss factor of substrate material; β_2 is the loss factor of damping material.

2.2 Practical Application and Improvement of Resonance Method

When using cantilever beam bending resonance method to measure dynamic mechanical parameters of damping materials, the measurement error comes from three aspects: first, the error caused by the measurement itself, such as reading error, instrument error and measurement method error; second, the theoretical error, such as the error caused by translation, is calculated by applying the theoretical formula; third, the error caused by the response of the measuring device to the drying of the measuring value.

According to Ma Shaopu's analysis of the above errors, the resonance frequency measurement error has the most significant influence on the calculation of loss factor and shear modulus[4]. Therefore, it is necessary to accurately control the parameters related to the resonance frequency measurement in the design of the actual test system, the sample design, the selection of machining precision and the allowable range of test error. Considering the influence of material damping of composite specimen, Hu Weiqiang[5,6] modified the measurement theory. The simulation results show that the base material with high damping should be avoided as much as possible in the design of samples with free damping structure. When the damping material has to be chosen, the improved measurement theory must be used to obtain more reasonable results. Wang Chao[7,8] applied two sided additional free structure damping specimens to the experimental subjects, and the concept of influence coefficient is proposed through sensitivity analysis. It is applied to analyze the influence of measurement error of parameters such as thickness ratio, resonance frequency ratio, density ratio and loss factor on the test precision of viscoelastic material. Furthermore, some important technical specifications are proposed for the design of specimen.

3 Half-power Bandwidth Method

3.1 Basic Principle

The damping properties of structures are tested by the half power bandwidth method, and it is simple and feasible and the precision can meet most engineering requirements, so this method has been widely used.

The half power bandwidth method is to calculate the damping ratio according to the influence curve of resonance of simple harmonic vibration system, especially under the circumstance of environmental excitation. The calculation formula is as follows:

$$D = \frac{f_2 - f_1}{2f_n} = \frac{\Delta f_w}{2f_n} \quad (3)$$

D——Damping ratio;
f_n——System resonance frequency;
H——Formant amplitude;
f_1, f_2——The amplitude is the two frequency points corresponding to 0.707H;
Δf_w——Half-power bandwidth.

3.2 Practical Application and Improvement

With cambered plate structure as the research object, Li Hui[9] set the hammer, vibration platform, three test systems of piezoelectric excitation, respectively using the method of half power bandwidth of damped free vibration attenuation method and time domain performance was tested. The study found that, through the shaking table sweep motivation method for vibration response, the method of half power bandwidth segmented fast fourier transform (FTT) after processing is used to identify the damping of the frequency response of the highest accuracy.

The half-power bandwidth method is used to obtain damping by using power spectrum. The advantage of this method is that it can be averaged multiple times. Liu Jinming[10] derived an accurate algorithm for solving damping without increasing sampling length and improving frequency resolution. According to the derived formula, the more accurate damping, frequency and amplitude can be obtained by least square fitting near the maximum peak point.

In addition to the above common methods, many researchers have translated measurement analysis methods that are more consistent with their own measurement structures when measuring loss factors.

Teng Tso-Liang[11] used the complex stiffness method to analyze and calculate the loss factor of the three-layer damping structure laminated beam; Mead[12] used the method of complex stiffness to calculate and measure the damping and the dissipation factor of the free damping beam and plate, and used it as an important method to estimate the damping structure; Gallimore[13] adopts the method of complex stiffness, which makes theoretical calculation and experimental verification for the dissipation factor of the restraint damping device use in the small aircraft ground floor system; Aenlle[14] used the complex stiffness method to analyze and calculate the loss factor of laminated glass damping structure; This paper proposes a new transition constraint damping structure, and verify the loss factor of the structure of expressions based on the complex stiffness method.

4 Conclusions

After analyzing the damping test method of many researchers, it can be obtained:

(1) Adjacent peak or trough can be collected in the freely attenuated waveform, or both peak and trough can be collected to eliminate the influence of direct flow. The time domain method is an accurate method in accordance with the damping ratio mathematical model, but it requires the signal to be a single-frequency free attenuation signal.

(2) There are many factors influencing the cantilever beam resonance method, so the size of the experimental components needs to be controlled.

(3) The half-power bandwidth method is greatly influenced by parameters. Once the parameter setting is unreasonable, the measurement result will deviate several times to dozens of times. In addition, there is a large error when the parameters are low damping and low frequency.

References

[1] Dai depei. Damping and noise reduction technology[M]. Xi,an: Xi,an Jiaotong University,2007.

[2] Xu Fengchen, Li Honglin, Liu Fu. Dynamic damping coefficient of the determination of damping materials [J]. Journal of adhesive, 2013, 9(3): 45-49.

[3] Garrett S L. Resonant acoustic determination of elastic moduli[J]. Journal of the Acoustical Society of America,1990,(88):210-221.

[4] Ma Shaopu, Wang Minqing, Hu Weiqiang, Zhang Junfeng, Liu Yansen. Theoretical error of resonance beam method for symmetric sandwich structures[J]. Noise and vibration control,2008(04):38-41.

[5] Hu Weiqiang, Wang Minqing, Liu Zhihong. Analysis on the impact of base damping on the measurement of the bending resonance of cantilever beam [J]. Vibration and impact, 2008(06):170-172+182+196.

[6] Hu Weiqiang, Wang Minqing, Sheng Meiping, Liu Zhihong. Research on broadband test of dynamic performance parameters of damping materials [J]. Mechanical science and technology,2007(11):1425-1428.

[7] Wang Chao, Lu Zhenhua, Gu Yeqing. The design method of the constraint damping test for the mechanical parameters of viscoelastic damping materials [J]. China mechanical engineering, 2016,27(09):1208-1214.

[8] Wang Chao, Lu Zhenhua. Mechanical parameter test of viscoelastic damping materials. Study on the design method of a bilaterally attached free structure damping specimen [J]. Vibration and impact,2014,33(05):102-108.

[9] Li hui, Sun wei, Zhang Yongfeng, Han Qingkai. Comparison of several test methods for the damping properties of cantilever plate structure [J]. Journal of Chinese engineering machinery,2013,11(04):347-353.

[10] Liu Jinming. Precise methods of using power spectrum for damping and frequency [A]. Proceedings of the 11th, 12th and 13th national vibration and noise high technology and application conference [C]. China vibration engineering society vibration and noise control professional committee,1999:6.

[11] Teng Tso-Liang, HU Ning-Kang. Analysis of damping characteristics for viscoelastic laminated beams [J]. Computer Methods in Applied Mechanics and Engineering, 2001, 190:3881-3892.

[12] Mead D J. The measurement of the loss factors of beams and plates with constrained and unconstrained damping layers: A critical assessment [J]. Journal of Sound and Vibration, 2007, 300 (3-5):744-762.

[13] Gallimore C A. Passive viscoelastic constrained layer damping application in small aircraft landing gear system [D]. Blacksurg V A: The Virginia Polytechnic Institute and State University,2008.

[14] Aenlle M L, Pelayo F. Frequency response of laminated glass elements: Analytical modeling and effective thickness[J]. Applied Mechanics Reviews, 2013, 65:1-13.

(13) Research Progress on Viscoelastic Damping Materials and Damping Structures

Wu Di

Institute of Functional Materials, Qingdao University of Technology

Abstract

Qtech viscoelastic damping material developed by functional materials research institute of Qingdao University of Technology has excellent damping properties. Based on Qtech damping materials, the influence factors of damping performance were studied, including constraint layer, damping layer thickness and environmental factors. The application of damping materials and damping structure in practical engineering has obvious effect of damping.

Key Words

Viscoelastic Damping, Constraint Layer

1 Introduction

There are many measures to reduce vibration and noise, but more and more engineering projects tend to use viscoelastic damping materials to make constrained damping structure to resolve the problem.

Viscoelastic damping material is a kind of polymer material. To enable the damping materials to be used in a wider range of temperature and frequency domains, Millar proposed the concept of interpenetrating polymer networks. Since then, many domestic experts and scholars have conducted a lot of researches on IPN modified damping materials, and increased material damping by adding an inorganic filler to the damping material.

At present, Qtech series viscoelastic damping materials have excellent performance of vibration and noise reduction due to their wide damping temperature range and high damping ratio. In addition, the influencing factors of damping property of damping structure are studied.

2 Qtech Viscoelastic Damping Material

2.1 Qtech501

The viscoelastic damping material Qtech501 consists of two components, A and B. Group A is divided into isocyanate, B components include amine terminated polyether (ATPE), chain extender and auxiliaries. The viscoelastic damping material Qtech501 is produced by a series of chemical reactions after components A and B collide with each other at high temperature and pressure. The equation is:

$$R-NCO+R-NH_2 \xrightarrow{\text{High temperature, High pressure}} R_2NHCONH$$

Qtech501 is analyzed by dynamic thermal mechanical analyzer, and the test results are shown in Figure 1.

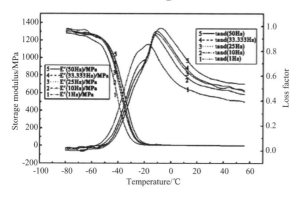

Figure 1　DMA results of viscoelastic damping materials Qtech501

As can be seen from Figure 1, Qtech501 viscoelastic damping material has a wide temperature field. Within the scope of the $-20\ \text{℃} \sim 60\ \text{℃}$, the energy storage modulus of Qtech501 viscoelastic damping material is less than 50MPa and the loss factor is greater than 0.4[1]. Under the influence of temperature, the change law of energy storage modulus and loss factor of the material is mainly determined by the relationship between the caloric value of molecules in material at different temperatures and the energy required for the motion of molecular segments[2].

2.2　Qtech503

Qtech503 also consists of two components, A and B. Component A: Diphenylmethane-4,4'-diisocyanate (MDI); Component B: amine-terminated polyether (D-2000).

Gao Jingang[3] studied the general and dynamic mechanical properties of Qtech503. The tensile strength of Qtech503 gradually increased with the time, and by the 20th day, it basically stabilized at 7.9 MPa. The elongation at break of Qtech503 gradually decreased with the time, and was basically stable at the 12d, at 344.56%. The gel time is 186.33s; The surface drying time was 10.67min; The hard drying time was 32.67min. The curing time of Qtech503 can not only guarantee the good adhesion performance of the substrate, but also save the construction time and obtain better economic benefits. As can be seen from Figure 2, the energy storage modulus of Qtech503 decreased with the increase of temperature. The material loss factor increases first and then decreases with the increase of temperature. The frequency mainly affects the position of energy storage modulus and loss factor curve of Qtech503.

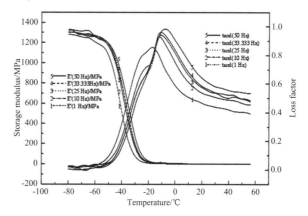

Figure 2　DMA results of viscoelastic damping materials Qtech503

3 Study on Vibration Performance of Damping Structure

3.1 Influencing Factors of Damping Property of Damping Structure

Most polymer damping materials have low elastic modulus and cannot be used as structural materials. They must be adhered to the damping member to form a composite damping structure. In a composite damping structure, rigid materials provide strength, and the damping materials transform mechanical energy into heat energy through deformation.

Huang Weibo[4] analyzed the influence of the thickness ratio of each layer and the thickness ratio between layers on the dynamic mechanical properties of the mortar sandwich plate structure. The research shows that the structure complex loss factor is the largest when the ratio of the constraint layer thickness to the base layer thickness is 1, and the ratio of the constraint layer thickness to the damping layer thickness has a greater influence on the structure composite loss factor. Lyu Ping[5] found that low temperature or high temperature environment can also deteriorate the damping performance of the constraint damping structure, but it is good at the room temperature. Yang Yang[6] studied the dynamic mechanical properties of sandwich plate damping structure under two supporting modes, including elastic support and fixed end constraints. And it is found that the thinner the damping layer is, the higher the efficiency is. Li Huayang[7] found that the local constraint damping structure with 80% of the laying area of the constraint damping layer has better damping effect than the structure with other laying areas. Lyu Ping[8] analyzed the damping property of the sandwich composite beam with the traditional epoxy resin and the new polyurea viscoelastic layer. The results show that the polyurea viscoelastic layer has good damping properties.

Based on the finite element analysis, Li Baojun[9] studied the damping performance of the restrained damping structure based on mortar board and steel bar. Huang Weibo[10] studied the vibration performance of the restrained damping structure of the strip interface. The results show that the loss factor of the stripe interface constraint damping structure increases with the decrease of the convex width of the bar interface and the increase of specific surface area. In addition, the influence of covering rate on the damping performance of partially restrained layer damping structure was studied by means of two-side simply supported single point impact test.

"Short circuit" refers to the phenomenon that the base layer and the constraint layer are not completely separated by the damping layer in the constraint damping structure, resulting in rigid contact. Li Dong[11] finds that with the decrease of "short circuit" area, the natural frequency of restrained damping structure decreases, and the loss factor increases.

3.2 Combined Application of Damping Structure and Practical Engineering

With the increasing requirement of vibration reduction in engineering, we also intensify the research on the combination of damping materials and damping structure with actual engineering.

Ju Jiahui[12] used dry and wet cycle experiments and high and low temperature cycle experiments to investigate the effect of ambient temperature and humidity on the performance of damping materials. And he used the road as the base and asphalt concrete as the constraining layer to form the vibration damping performance of the pavement-constrained damping structure. Lyu Ping[13] expounded the feasibility and application prospect of pure polyurea as a new type of protective material for damper type bridge support from the aspects of protective performance, aging and damping performance, construction technology and existing engineering application. And used Ansys to establish the plane finite element model of soil-tunnel-track bed, and analyzed the vibration response of the integral track bed and the viscoelastic damping track bed in the tunnel structure under the action of frequency load of 5~25 Hz. Liang Longqiang[14] found that it is helpful for vibration and noise reduction in metro operation process to spray a layer of 2~3 mm thick Qtech viscoelastic damping material between the concrete subgrade and ballast bed. Huang Weibo[15-17] designed three kinds of constrained damping bed for subway vibration and noise reduction, which are respectively serrated interface, strip interface and grid interface.

The damping property and waterproof property of the damping materials are studied with the metro project. The vibration response of subway station, elevated track and the ground under the beam is tested.

4 Future Development Direction

(1) The elastic modulus of the constrained layer is increased by the aggregate and the effect on the damping performance of the constrained damping structure is studied.

(2) Through the dynamic response test, the modal analysis of the damping structure is carried out to study the modes of the various stages of the damper structure, which provides a certain basis for optimizing the design of vibration reduction.

(3) The dynamic mechanical properties of viscoelastic damping materials are

affected by ambient temperature. Mechanical properties at other temperatures are further studied.

References

[1] Lyu Ping, Gao Jingang, Li Jing, Zhang Zhichao. Research of Damping Performance of Constraint Layer Damping Based on Qtech 501 Viscoelastic Damping Material, Industrial Construction, 2014. 44:354-357.

[2] Huang Weibo, Gao Jingang, Li Jing. Research of Damping Property of Viscoelastic Damping Materials 501 and Structure, Development and Application of Materials, 2014. 29(04): 82-85.

[3] Gao Jingang. Study on Performance of Damping Materials Qtech 503 and Constrained Damping Structure in Qingdao Subway. 2014.

[4] Huang Weibo, Zhang Zhichao, Li Huayang, Ma Yanxuan. Study on the Effect of Layer Thickness on the Vibration Property of Constrain Layer Damping Structure. Earthquake Resistant Engineering and Retrofitting, 2018,40(01):8-14.

[5] Lyu Ping, Gao Jingang, Li Jing, Bo Zhongwei. Impacting Factors of Damping Performance of Constraint Damping Structures. Noise and Vibration Control, 2014, 34 (05): 234-238.

[6] Yang yang. Study on the Viscoelastic Materials and Dynamic Properties of Viscoelastic Laminated Structure. 2016.

[7] Li Huayang, 2018. Study on the Viscoelastic Materials and Dynamic Properties of Partial Constrained Layer Damping Structure. 2018.

[8] Lyu Ping, Liu Xudong, Ma Xueqiang, Huang Weibo. Analysis of damping characteristics for sandwich beams with a polyurea viscoelastic layer. Advanced Materials Research, 2012, 374-377: 764-769.

[9] Li Baojun. Research of a Constrained Damping Structure Based on a New Type of Viscoelastic Material and ANSYS Analysis. 2011.

[10] Huang Weibo, Li Dong, Feng Chao, Liu Tiancheng, Ju Tao. Study on Vibration Characterize of Strip Interface Constrained Damping Structure Based on Finite Element Analysis. Industrial Construction,2017, 47:111-114.

[11] Li Dong, Huang Weibo, Liu Tiancheng, Li Xiangdong, Ju Tao. Study on the Influence of Short-circuiting on Damping performance Based on Finite Element Analysis. Engineering Construction, 2017, 49(07): 18-22.

[12] Ju Jiahui. Experimental Research on Viscoelastic Damping Material and Damping Performance of Pave-

ment Constraint Damping Structures. 2018.

[13] Lyu Ping, He Xiaoshan, Chen Kaihua, He Xin, Huang Weibo. Research Progress of New Protective Materials for Damping Bridge Bearings. Materials Review, 2016,30(01):96-101.

[14] Liang Longqiang, Huang Jian, Huang Weibo, Yang Lin. Construction Technology and Quality Control of Spraying Viscoelastic Damping Material in Qingdao Metro. New Chemical Materials, 2017,45(01):228-230.

[15] Huang Weibo. The serrated interface constraint damper track bed used for subway vibration and noise reduction and its preparation method: CN, 106812029B, 2018-04-10.

[16] Huang Weibo. Banding interface restrained damping bed used for metro vibration and noise reduction and its preparation method: CN, 106592343B, 2018-02-06.

[17] Huang Weibo. Grid interface constraint damping bed for metro vibration and noise reduction and its preparation method: CN, 106835860B, 2018-04-10.

(14) Water Capillary Absorption of Alkali Slag Concrete after Salt-frost Action

Han Xukang, Wan Xiaomei, Yang Baoxian, Gao Caobo, Zheng Heping
School of Civil Engineering, Qingdao University of Technology

Abstract

In this paper, the deterioration damage and water capillary absorption of concrete after frost action based on alkali-activated cementitious material are researched experimentally. From the experiment, the damage degree of alkali-activated cementitious material concrete is studied in combination with the microscopic test. Through chloride and freeze-thaw cycle test, and water capillary absorption of specimens after freeze-thaw cycles in chloride solution are researched as well. Based on the results, the effect of mix proportion especially activator type on frost damage degree and water transport property are discussed.

Key Words

Water Capillary, Alkali Activated, Slag Concrete

1 Introduction

Compared with the traditional Portland cement concrete, it is generally believed that alkali-activated concrete has better long-term durability. Frost heaving cracking and surface shedding are the two most important forms of concrete damage caused by frost. In the study of durability related to frost damage, Gifford[1] and other studies pointed out that after a certain number of freeze-thaw cycle, the alkali-activated cementitious material had the same mass loss rate as the ordinary Portland cement. Cyr[2] pointed out that compared with the ordinary Portland cement concrete, the alkali-activated concrete material could resist many times of freeze-thaw cycles, while the internal damage and external exfoliation and the mass loss are small.

The impermeability directly affects the frost resistance of concrete. Therefore, the basic research on the impermeability of alkali slag concrete is of great significance. The purpose of this paper is to study the permeability of alkali slag concrete after freezing. Through the freeze-thaw cycles in chloride solution, the permeability of alkali slag concrete after freezing is analyzed, and the frost resistance is studied which is an important part for the study of its durability.

2 Raw Materials and Experimental

2.1 Raw Materials

Main raw materials of the test include slag, grade two fly ash, sand, gravel, sodium silicate and sodium hydroxide solution. The chemical composition was determined by X fluorescence analysis as shown in Table 1. The slag used in this experiment comes from Hao De Company, while fly ash comes from Qingjian group.

Table 1 Chemical composition of slag and fly ash, %

%wt	CaO	SiO$_2$	Al$_2$O$_3$	MgO	Fe$_2$O$_3$	SO$_3$	Na$_2$O
Slag	41.6	26.8	17.79	9.28	0.48	2.03	0.31
fly ash	5.79	66.8	17.93	1.5	4.03	0.5	0.26
%wt	K$_2$O	MnO	SrO	BaO	P$_2$O$_5$	TiO$_2$	I.R.
Slag	0.39	0.34	0.11	0.09	0.02	0.72	0.03
fly ash	1.32	0.06	0.08	0.08	0.43	1.07	0.15

2.2 Experimental

2.2.1 Mix Proportion

In the early stage, a large number of experiments and orthogonal experiments were carried out to determine the most suitable mix proportion (Table 2、Table 3).

Table 2 Main parameters of alkali activated slag concrete

Group	Activator	Alkali equivalent (Na$_2$O)	Solution (Kg/m^3)
1	NaOH	7%	184
2	NaOH	9%	184
3	NaOH	9%	184
4	Na$_2$SiO$_3 \cdot$ 9H$_2$O	7%	184
5	Na$_2$SiO$_3 \cdot$ 9H$_2$O	9%	184
6	Na$_2$SiO$_3 \cdot$ 9H$_2$O	9%	184
7	Na$_2$SiO$_3 \cdot$ 9H$_2$O	9%	184

Table 3 Main parameters of alkali activated slag concrete

Group	Gravel (Kg/m^3)	Sand (Kg/m^3)	Mineral powder (Kg/m^3)	Fly ash (Kg/m^3)
1	1074	716	400	0
2	1074	716	400	0
3	1074	716	320	80
4	1074	716	400	0
5	1074	716	400	0
6	1074	716	320	80
7	1074	716	400	0

2.2.2 Freeze-thaw Cycles in Chloride Solution

In the salt freezing test, the fast freezing method was adopted. The cured 100 mm^3 cubic specimens were taken out after the freeze-thaw of cycles in chlorine solution, and the change of the dynamic modulus and the loss of the mass were measured (According to GB/T 50082-2009 The standard of test method for long term performance and durability of ordinary concrete).

2.2.3 Capillary Water Absorption Test

Capillary water absorption tests were carried out after freeze-thaw cycle of the samples. The 100 mm^3 cubic specimens were removed after 125, 175, 225 cycles. After drying to constant weight, it was put into the water tank for water capillary absorption test.

3 Test Results and Analysis

3.1 Relative Modulus of Kinetic Elastic Modulus After Salt Frost Test

After analyzing and sorting out the preliminary data of salt frost test, the relative dynamic modulus diagram of the specimens was obtained, as shown in Figure 1. The change of relative dynamic modulus was relatively fast at the early stage of salt frost test. With the freeze-thaw cycle, the surface of the specimens is frozen, the hydration product of the inner layer is more stable, and the inner salt concentration is relatively low and the saturation is lower, so the frost damage rate of the specimens is relatively slow and the relative dynamic modulus is relatively slow. Especially, the performance of NaOH as an activator in the experimental data was obvious.

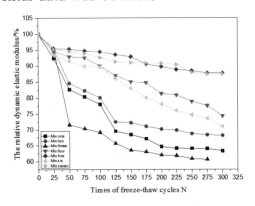

Figure 1 relative dynamic modulus after salt freezing test

The alkali slag concrete has good frost resistance. In the process of several times of freeze-thaw cycle, the decreasing trend of relative dynamic modulus is slow, and it can withstand more than 300 times of freeze-thaw cycle, and the relative dynamic modulus is still less than 60%, which is far superior to ordinary concrete.

3.2 Mass Loss Rate After Salt Freezing Test

Analysis of test data showed that the mass loss rate of concrete blocks varied with freeze-thaw cycles of salt frost test, as shown in Figure 2. From the data of each group, the effect of alkali equivalent, fly ash, modulus and activator on the frost resistance of concrete in the salt frost test was basically consistent with that of the water frost test. Compared with NaOH, sodium silicate was more beneficial to the frost resistance of concrete as an activator, and the alkali equivalent content and modulus had positive influence on the frost resistance of concrete, and the fly ash had adverse effect on the frost resistance of concrete.

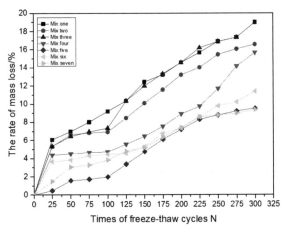

Figure 2 mass loss rate after salt freezing test

3.3 Analysis of Capillary Water

From the data analysis of each group in Figure 3, it was found that the capillary water absorption curve of the salt frost test was basically the same as that of the water. The capillary water absorption rate of the concrete with sodium silicate as an activator was slower, the structural damage was slower, and the frost resistance was better. The alkali equivalent content had a positive effect on the frost resistance of concrete, and the larger the alkali equivalent was, the better the frost resistance was. The mixing of pulverized coal ash had negative effect on the frost resistance of alkali activated slag concrete.

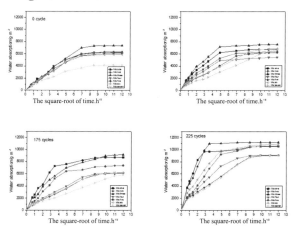

Figure 3 water capillary absorption curve after salt freezing test

4 Conclusions

Based on the results of the experiments, the following conclusions can be drawn:

(1) Alkali-activated slag concrete with sodium silicate as activator has better frost resistance and strength than alkali-activated slag concrete with NaOH as activator.

(2) Alkali equivalent content has a positive effect on the frost resistance of alkali-activated slag concrete.

(3) Within a certain range, modulus of sodium silicate is beneficial to alkali-activated slag concrete.

(4) The incorporation of fly ash is unfavorable to the frost resistance of alkali-activated slag concrete.

(5) The concrete with sodium silicate as activator has better frost resistance and impermeability, while the concrete with NaOH as activator has relatively low frost resistance and impermeability.

References

[1] Gifford P M, Gillott J E. Freeze-thaw durability of activated blast furnace slag cement concrete[J]. ACI Materials Journal, 1996, 93(3):242-245.

[2] Cyr M, Pouhet R. The frost resistance of alkali-activated cement-based binders[M]. Handbook of Alkali-Activated Cements, Mortars and Concretes, 2015:293-318.